B E Y O N D
GLOBAL
WARMING
THE BIGGER PROBLEM AND REAL CRISIS

BEYOND GLOBAL WARMING

THE BIGGER PROBLEM AND REAL CRISIS

JOHN DURBIN HUSHER

LitPrime
"Your story is our priority"

LitPrime Solutions
21250 Hawthorne Blvd
Suite 500, Torrance, CA 90503
www.litprime.com
Phone: 1-800-981-9893

Published by LitPrime Solutions   08/11/2021

ISBN: 978-1-954886-66-7(sc)
ISBN: 978-1-954886-67-4(hc)
ISBN: 978-1-954886-68-1(e)

Library of Congress Control Number: 2021916736

CONTENTS

DEDICATION

To the engineers and scientists of the world and
their solutions for tomorrow's problems

PREFACE

ecently, numerous articles have been written about the possibility of Earth's global warming. Because the Sun has been the constant and sole provider of Earth's energy for at least the last billion years, I wondered what changed to make this problem occur. Every year, the Sun provides Earth 5.5 x 10^{24} joules (watts) of energy. This, the solar constant, which has been measured and confirmed by satellite measurements. The Sun's energy provides us with the following major functions. It warms land and water, provides a water cycle that provides the world's fresh water, and provides wind and ocean currents that spreads the extraordinary heat of the equator towards the North and South Pole to normalize the temperature across Earth. Additionally, a small portion of this energy provides photosynthesis, the basis for plant growth and our food.

Because the energy being supplied is constant and energy cannot be created or destroyed, I wondered what changed to possibly cause global warming. It takes energy to produce increased heat and result in global warming. If energy is going to be used to cause global warming and we have constant energy coming from the sun, then something on our planet has to change and give up some of the energy it has been enjoying. At least this was my first glance at the subject matter.

With these thoughts in mind, I decided to review the suggested problem of global warming. In order to make it easier for the reader to comprehend the various energies described, I have converted all energies to Calories; that is, the table Calories that one can easily relate to. The

Sun's energy, the weather's energy, and all other energies are described in units of Calories. For example the solar constant mentioned in the first paragraph is equal to 1.315 x 10^{21} Calories (food Calories).

In order to do justice to the subject matter, I went back in history and started there. What has been the history of Earth and its climate changes? What were the causes? I decided to review three phases of Earth's existence: before man, after man, and after oil. The reason for this three-phase approach was to determine how Earth, without man or oil, handled or provided climate changes as well as determine if global warming is one of those changes.

Some changes came with the arrival of man. What if the increase in the number of people was indeed a potential cause? Not too long ago, (300BC) about 100 million people were on this planet. Today, approximately 6.5 billion people are here. Going from 100 million to 6.5 billion means the number of people has increased by a factor of sixty-five. Isn't that a huge number? Just think of 6.5 billion people walking around with a temperature of 98.6 degrees Fahrenheit (37 degrees Celsius) and breathing in oxygen and breathing out carbon dioxide. That's a lot of heat and carbon dioxide. Human beings generate heat and carbon dioxide as they walk around and produce other things that generate heat and carbon dioxide. Could our existence on this planet cause the suggested global warming?

For the third phase, I reviewed when oil and the fossil fuels became a major part of Earth's activities and if this related to the suggested global warming phenomenon. The Oil provides us energy. But where did it come from? Here is an earthborn energy separate from our Sun's budget to Earth. Here was earthborn energy that became a major part of our energy consumption in 1859.

This review showed that Earth's history contained many climate changes before man and before oil. I detail a reason for each. This story alone is worth reading for those of you who are unaware of the various challenges that Earth faced before we arrived at our present state of living.

With man and animals increasing to our present large numbers, I felt this would have a major impact on the energy that the Sun supplied.

Man consumes food and water that is derived from the Sun's energy. I reviewed how many table Calories, a measurement of energy, that a man consumes every day and the total for 6.5 billion people. In addition, I reviewed the table Calories that animals consume to provide man's other source of food. Was the energy being consumed anywhere near the energy being supplied each day by the Sun? Did this have an impact? Even though this added up to an enormous amount of energy, it did not explain the issue. As I reviewed these numbers, it became fairly obvious that the Sun only has the following indirect impacts on man; it keeps Earth at a temperature that is livable, and it supplies the sunlight for the plants that man consumes as well as those consumed by animals that man eventually takes as food. The Sun provides fresh water for man, animals, and plants. Besides the need for this food and water provided by our sunlight, man could live in the dark and survive.

This review proved quite rewarding. The energy to supply man across the planet burns up considerable energy; leading to a major problem that exists *beyond global warming*. So I continued my pursuit for determining where the balance might have changed. I began a review of water vapor because it is a higher quantity greenhouse gas than carbon dioxide. There is much more water vapor than carbon dioxide in our atmosphere. I pursued this avenue for an answer. The answer could be in the water vapor. Actually, until now, water vapor has been seen as our savior, not culprit.

Because my review started early in the life of Earth, I could see some other possible culprits due to events that occurred many years ago. Sixty-five million years ago, a meteor hit Earth with a tremendous amount of energy. Earth had to assume this energy. As a result, dust filled the atmosphere, causing the demise of the dinosaurs that inhabited Earth. The accumulation of dust covering these large, dead creatures resulted in the burial of a form of unusable energy at the time. I term this unusable energy as entropy.

In 1859, the first commercial oil well was dug in Titusville Pennsylvania. This oil, the remains of those dinosaurs, and other material, had been buried for sixty-five million years. Now it was providing the extra energy needed for the suggested global warming.

Here's where the extra energy, other than the sun, came from. But did it add to any global warming problems? The answers are here.

This third phase of the review of history became a revelation for me and possibly for those of you who read this account. Partially due to the meteor and other material buried for years, I found the extra energy of entropy I was looking for. Besides the Sun's constant supply and some small meteorites that enter our atmosphere, no outside energy has been coming into our planet. Here, buried for millions and millions of years, was a terrific amount of extra energy. We have been using some form of this energy over the last 150 years. As benefactors, the world has changed since the industrial revolution and the use of fossil fuels for energy.

You will find that the suggested global warming can not be blamed on the automobile exhaust as some have indicated. There are other phenomena that generate much more carbon dioxide than automobile exhausts. As much as I am happy to see Mr. Al Gore receive the Noble Peace Prize, I will show he is wrong in his pointing towards automobile exhausts. I am happy for him, since he brings this issue to the forefront and has worked long and hard on this subject matter. I wouldn't have worked on it if he hadn't brought it to the forefront.

As I pursued this approach, I decided to quantify the energy we presently use, the growth expected, and the amount in the world's reserves. I arrived head-on into another bigger problem than the suggested global warming. The one I discuss is real. It does not take a scientist to analyze. The main purpose of this book is to bring this major issue to the forefront, since it is an issue that can be overcome with a concentrated worldwide effort.

In addition to absorbing the study on Earth's energy, you will be interested in reading the first phase which reveals that this is Earth's third atmosphere and what happened to the other two. Earth had super global warming in its early phases, and it relates to the cause today. You will wonder at Mother Nature's tricks in transformation to overcome many significant issues. Not long ago, something happened to change the world we inhabit. You will find the answer to the global warmth we enjoy today as well as the present possibility of global warming. More importantly, you will find something that is *beyond global warming.* It is something that will become obvious to you as you read through and understand why it is more important and more immediate than global warming.

You will be interested in my suggestions for solving this new issue with actions to be taken in the not too distant future. I am hopeful that many of you will come up with other ideas to avert this issue. Man does have a chance to overcome the consequences of this issue if enough focus and intelligence is placed on the solutions.

I have spent most of my life as an electrical engineer and an industry leader who solves problems. For readers who are not engineers or otherwise scientifically trained, don't worry about this text being too technical. I have written this in a manner that skips the equations and goes forward with common sense and a wonderful story about Mother Earth and presents this situation that is *Beyond Global Warming.*

Just be patient as we go back to the big bang and see what history tells us about Earth and its ability to cope with problems. Mother Nature has a solution for global warming.

INTRODUCTION

O ver the past several years, I have read many articles where scientists predicted that global warming is now upon us. In the late 1970s, during some years of cold weather, the papers quoted several scientists who claimed that Earth was about to have a global ice period. This notion soon dissolved in the wake of the warmer weather years that followed.

Many of these same scientists did not predict global warming in the 1990s through 2005, even though 1998 was the warmest year in the history of recorded global temperatures. However, because 2005 was another warm year that might have exceeded 1998, an increased breakout of predictions about global warming resulted. The main theme related to the assumption that most of the problem was due to automobile exhausts and their related carbon exhaust.

Based on previous forecasts, I wondered if scientists could make an accurate estimate of Earth's warming and how it would impact all of us. Maybe they picked this subject, like I did, because it was popular or frightening. Consequently, they gained a larger audience. It is also a popular subject because the environmentalists picked up on it right away. In fact, they may be the ones who started it.

I then decided to pursue a more detailed review of global warming. In order to do justice to the subject matter, I would review history and determine if there had been any global warming before man. I also wanted to evaluate any climate changes that might explain the rise in temperature that we may be experiencing now and how the planet has responded to these problems. I knew the history of Earth was one of major changes. I also knew that man, animals, and cars were not around to complicate the picture back then.

The review showed there had been global warming and many climate changes in Earth's history before man. A reason for each was found. I will discuss these and work my way up to the present. You will see what impact man had on climate as well as how fossil fuels entered the picture. What is so different between the past and the present? Why is there an abrupt change? Or, is there? Is a change really coming for the people of Earth? I will answer these questions. There are surprises for those of us who have not looked back and then moved up to the present. I was shocked to find another problem that was worse than this global warming issue. It is real and does not take a scientist to analyze.

In order to make it easier to relate to the various energy terms, I used the Calorie as the unit of power. This is the same Calorie that we think about while we are dieting. It is the same Calorie as the table Calorie. It is not the small calorie that scientists and engineers use. The table Calorie uses a capital C since it is equal to the kilocalorie (1000 small calories) that scientists and engineers use. Wherever joules or watts were used, I converted these to Calories to provide a constant picture of energy use. I also describe the Sun's energy in Calories.

Before reading this book, review the index in the back of the book. Learn about the various forms of energy as well as the units that quantify the energy. This index talks about BTUs, Therms, and kilowatt-hours that each of you see in your monthly gas and electric bill. I took these terms and their units and converted them to Calories to make it easier for you. Separate from the main subject of global warming, you will be interested to learn that this is Earth's third atmosphere. I will also discuss about what happened to the other two. You will be amazed at Mother Nature's tricks. Earth goes from a crude planet to one of a polished nature. It is like the metamorphous of the worm into a butterfly.

Oddly, today's scientists are concentrating on carbon dioxide when water vapor might actually contribute more to any global warming that may be coming. Water vapor, or precipitation, will serve in a double role of being the culprit or a redeemer that could counteract the effects of carbon dioxide and any global warming.

Not too long ago in our planet's history, an event occurred that is presently providing us a double-edged sword: the good, the bad, and maybe the ugly. I have analyzed the situation based on the concept that energy cannot be created or destroyed in a controlled (closed) system. Using this analysis, I paint a picture of Earth as we find it today and where we have to take it tomorrow. Partnering with Earth and its massive contained energy, man will be able to derive some interesting solutions to problems that will be upon us in the near future. There are answers to these problems that are *beyond global warming*. Eventually, global warming will be a secondary issue. I will provide suggestions for the present and the future. Man does have a chance to evade the consequences of the future. But the future is not the one we are looking at. It is one that is rapidly descending upon us now. You will find it interesting to stop and think about various ways out of this dilemma. I have provided some solutions, but each of you will be in a position to have your own ideas and suggestions.

The Present (Before the Analysis)

Over the past several years, I have read many articles about the possibility of global warming. I have spent a great deal of time researching a possible cause. I have had several different reactions due to the dichotomy of current scientific and political theory. Many countries have signed the Kyoto Protocol, which is predicated on reducing carbon dioxide that is dumped into the atmosphere. But the United States has not signed the agreement. At times, I felt we did not sign because of the impact it might have on our economy. We are the biggest producer in the world of consumer items. We are also the biggest producer of carbon dioxide that is dumped into the environment.

Are we the only bad guy in this scenario? Will China soon pass us by and be the leader in this regard? Are these scientists correct in their predictions of a large increase in global temperature and the increase in the water levels around the world?

Maybe I am hard to convince because weather forecasters cannot provide an accurate weather forecast for two weeks down the road. I realize there is a difference between weather forecasters and the scientists that forecast climate change. But, in both cases, they are working with Mother Nature; and Mother Nature has a way of confusing even the best of us.

If you decide to pick a related subject, like air pollution by automobiles, as the framework for your analysis, global warming is a fairly easy subject to write about. It is easy for you to read about how automobiles are polluting the atmosphere with carbon dioxide and believe this to be the main culprit. Further, it is easy for you to read that carbon dioxide forms a barrier that prevents radiation from Earth's surface from escaping into the atmosphere and space above, thus resulting in global warming. You, an average reader, can close your eyes and envision this happening. So can I. I am not an automobile advocate. I do believe that it is possible that the automobile could be the culprit if there is warming.

However, I can also close my eyes and envision other culprits. For example, the Earth's current human population could be the culprit. Not too long ago, only about a billion people were on this planet. In fact, one would not have to go far back in history to find a time when only about 100 million people were on Earth, just before the beginning of the Common Era. Now, there are just over 6.5 billion people are on this planet.

To go from 100 million to 6.5 billion means the number of people has increased by a factor of sixty-five. That is a huge increase. Just think of 6.5 billion people walking around with a temperature of 98.6 degrees Fahrenheit (37 degrees Celsius). They are breathing in oxygen and breathing out carbon dioxide. That's a lot of heat and a lot of carbon dioxide.

Human beings generate heat and carbon dioxide when they walk around. They also produce other things that generate heat and carbon dioxide. Maybe we are the culprits. I review how we contribute to global warming and other phenomena that should interest you. Maybe both people and automobiles are the culprits.

On the other hand, we may be going through a climate change. Not too long ago, we went through an ice age. But history has shown that large areas called carbon sinks can handle carbon dioxide. These relate to the plant life and forests of our world as well as the sea plants that provide very large sinks for taking on carbon dioxide in the oceans of the world. This also relates to these same areas and their ability to generate oxygen.

This subject of generators and sinks relates to a balance, —a balance between what generates oxygen (plants on land and in the ocean) and carbon dioxide (animals and humans) and what absorbs oxygen (animals and human beings) and carbon dioxide (plants on land and in the ocean). There are other generators of oxygen. For example, weathering releases oxygen from rocks and minerals. How well this balance is sustained may determine if we have the essential elements to offset global warming, assuming global warming exists and is caused by the phenomenon of too much carbon dioxide in the atmosphere.

First, we must examine the reasons why scientists contend that the generation of carbon dioxide is the main contributor to the greenhouse effect. Earth's atmosphere consists of the following:

Gas	Percentage in atmosphere
Nitrogen	78.1 percent
Oxygen	21.0 percent
Carbon dioxide	0.038 percent (380 parts per million)
Methane	0.0015 percent (15 parts per million)
Water vapor	0.5 to 1.0 percent (approximately 5 to 10 thousand parts per million) *
Various other gases make up the remainder of the atmosphere (approximately 0.4 percent)	

Table 1. Breakdown of Earth's atmosphere.
* Percentages of all gases changes as water vapor content changes

The main gases that contribute to the greenhouse effect are carbon dioxide, methane, and water vapor. Water vapor generally varies the widest in parts per million and contributes the most to the greenhouse effect. (40) Scientists contend that carbon dioxide is being generated at increasing levels and may soon be the major contributor to the greenhouse effect. They base this statement on the increased burning of fossil fuels through the use of automobiles and various other human endeavors and the jump in atmospheric carbon dioxide over the last sixty years.[1] It went from approximately 280 parts per million to 380 parts per million. I will discuss this change later.

But it is important to review these percentages and realize that water vapor is running from 5,000 to 10,000 parts per million, which is over an order of magnitude higher than carbon dioxide. If there is any global warming, water vapor may be the major player. Scientists had previously reviewed this subject on the amount of carbon dioxide in the atmosphere. They estimated that, without any automobiles and people, 95 percent of the carbon dioxide would be at its present level in our atmosphere. This is based on both very old data before people were present and recent data that determined that most of the carbon dioxide in the atmosphere is generated by wood rotting in dense forests due to the inability of the Sun to reach the trees and branches that have fallen in the major wooded areas of the world. Interestingly, young trees and plants take in carbon dioxide and release oxygen through the act of respiration. But, as these trees and plants become more mature, they generate carbon dioxide under certain conditions. The older forests fit into this category mainly because of their density.[2]

Most of the wood that is rotting occurs at various older tree lines that surround Earth in several major places. The one that produces 30 percent of the carbon dioxide in the atmosphere is the tree line in the upper latitude of Canada. This is the area, when viewed from the North Pole, where there are no trees while traveling south until one reaches the

---

[1] It increased from approximately 280 to 380 parts per million.

[2] En.Wikipedia.org/wiki/Carbon_dioxide_sink; P. Falkowski and others, "The global carbon cycle: a test of our knowledge of Earth as a system," *Science* 290 (2000): 191–296.

northern part of Canada. One then encounters a huge band of forests that stretches around the globe at this latitude. It represents 30 percent of the origin of the atmosphere's carbon dioxide. The trees are so dense that the Sun cannot reach the trees or branches that fall on the forest floor, which lay there to rot. As they rot, rather than being oxidized, they are reduced, thus forming a large differential in carbon dioxide.

Russia has the largest amount of forests that fall into this category. Located in Siberia, their location makes it difficult to harvest and transport them economically. They, therefore, represent old forest growth and will probably remain in this condition for years to come.

Another band of forest growth is located near the equator. The Brazilian rain forest is the largest of this type in the world. The second largest rain forest is located along the Congo in Africa. Although void of the large, stately trees of the northern area of the world, the vegetation is so thick and heavy that they also have a similar problem as the one just described in the northern latitudes.

Another large forest that emits large amounts of carbon dioxide is located in the northwestern part of the United States. Programs have been initiated to plant trees to replace the ones being harvested for the building of homes. This is a great program that should be extended to many other places around the world. These new growths represent areas where carbon dioxide is pulled from the environment and oxygen is generated. This is a very important program, even if carbon dioxide is not the problem portrayed.

Meanwhile, forests are being stripped all over the world to provide wood to build homes and burn for fuel. The removal of the world's forests proves to be an enigma. We want the forests to remain so they can provide the needed absorption of carbon dioxide. However, if the forests are too dense, they generate carbon dioxide rather than oxygen. Selectively cutting down the forests becomes a difficult decision, but one that must be made. Cutting down huge amounts of trees represents the sole income for many people of the world. Additionally, these forests are located in areas that are difficult to sustain other forms of agriculture that would provide another form of income for the local populace.

In addition to forest areas, jungles are located mostly in the Southern

Hemisphere. These may or may not be disrupting the balance between oxygen and carbon dioxide. The Southern Hemisphere has four times more water than land. Therefore, most of the vegetation that generates oxygen is in the oceans. These masses of ocean algae provide a huge sink for carbon dioxide. In the Northern Hemisphere, there is one-and-a-half times more water than land. The Northern Hemisphere does not have as great an ocean sink for carbon dioxide. At the same time, the Northern Hemisphere generates more carbon due to large industry and cars.

My Visual Impressions

While flying over most of the world and looking out the windows of the planes, it is interesting to see how much of Earth is still uninhabited and full of greenery. The only places that are sparse of greenery are the following:

- Large- and medium-sized cities (which generate carbon dioxide)
- Large desert areas (which are probably carbon dioxide neutral)
- Oceans (which are presumably active carbon sinks)

Irrigation programs, like the one that took Arizona from a desert area and made it a flourishing area, should be reviewed to determine if it is feasible to accomplish this feat in other areas of the world. Maybe, by judicious use of irrigation, other major areas could be made arid and support the sinking of carbon dioxide. The main issue would be obtaining water for the irrigation. Much of Earth's source of fresh water pours into the salty seas of the world. Programs for dams like the ones constructed in the 1930s should be reexamined as a source of irrigation.

The other great sinks of carbon dioxide are the oceans. Some review of the oceans and what could be done to increase the level of their ability to sink carbon dioxide has been done. Increasing the iron content in the oceans would make them more productive in this respect. However, this is not easily accommodated. And, with an increase of airborne carbon dioxide, photosynthesis would increase. Sea plants would grow more rapidly and offset this increase. In addition to the removal of the atmosphere's carbon dioxide, these sea plants would cause more oxygen to be generated in our atmosphere and oceans. Studies are taking place to determine the state of the world's oceans in relation to this problem and solution.

Maybe the older Forest areas are the major blame and solution of the carbon dioxide issue. However, these forest areas have been doing this since the beginning of our time on Earth. So, if there is global warming, these areas may be contributing to it. The increase in the number of people and automobiles has just added to it and taken it over the top. Maybe that's the scope of the problem, if there is a problem. We know Earth has gone through many cycles of climate changes in the past. Maybe this is one of those cycles. However, maybe this cycle is going to last so long that it will not be tolerable. Undoubtedly, to date, the papers written on global warming are frightening. However, the predicted effects are not so much for our generation. Perhaps our children or grandchildren will not be able to escape its effects.

Conservation of Energy

With this in mind, I decided to look into the subject in a different manner than others have. I am an electrical engineer. I am also an industry leader, having managed large electronic divisions in well-known companies over the years. In addition, I have two patents on integrated solar cell devices, which provide energy without any air pollution. These experiences do not directly qualify me to be an expert on the subject of global warming, but they do qualify me as a technical problem-solver. So, if global warming is a problem, why not just approach it as a problem to solve? A good engineer does not try to solve the symptoms of a problem. Instead, he or she tries to find the cause or source of the problem. In this case, global warming may just be a symptom, not the problem.

Maybe the number of people and automobiles are the problem. They may cause global warming, if there is global warming and it is not just a cyclic change in climate. Therefore, I decided to approach the problem like an engineer would. I would keep an open mind while studying the symptoms and problem areas. Then I could see what I could derive. I will study the problem of global warming, utilize the theory of conservation of energy that states, "Energy cannot be created or destroyed in a closed system," and see if this approach helps me to determine if there is a problem and determine the cause.

In order to pursue the problem in this manner, I must review history. I really mean going way back, back to where Earth was formed. I will review Earth's beginning from a macroviewpoint, devoid of man and automobiles. I will then succinctly bring us up to the last million years and set the stage for the coming of man and see if this brings me the information I am looking for.

After the initial phase of reviewing Earth's history, devoid of man and oil, I will go into a second phase that includes man. After injecting man into my studies and determining what effect he has had on the subject matter, I will begin a third phase and include the fossil fuels. I will analyze any effect they may have introduced in the current state of our atmosphere. With this three-phase approach, I am able to eliminate huge prime movers to ascertain information free of any contributions they may have made. I should be able to determine when and if the problem was interjected into our world. I would expect to see the various changes that the fossil fuels may have introduced and if they are the catalyst that could be initiating today's possible global warming. With this three-phase approach, I've been able to analyze the information more clearly and avoid many of the issues that could be contributed by the variables left out. After this review, we should be able to determine whether we are seeing a climate change or an instigated change. Mother Nature has provided many surprises in the past. Maybe she is at it again.

So, hang on to your hats as we go back to the big bang and see what history tells us about Earth and its ability to cope with problems void of man and fossil fuels. Mother Nature may have a solution up her sleeve that relates to global warming.

PHASE ONE

The Beginning and Stabilization of Earth

The Big Bang

One scenario for how Earth and the rest of the solar system and all the stars in space were formed is called the big bang. Essentially, an incredibly small physical source of all this energy exploded in the big bang and shot out material into space in all directions. The portion that represents our solar system is believed to have been a rotating cloud of solar dust and gas that began condensing into what is now the Sun, planets, and other solar elements, such as asteroids, meteorites, comets, and other solar junk. As this system originated, Earth was much bigger than it is now. And the Sun was not shining.

About 4.5 billion years ago, Earth began to reduce in size as gravity pulled the larger and denser parts of our Earth toward the center core. The Sun began shining, providing solar heat and solar winds, as stars do. During this period, Earth began to form an atmosphere comprised of the lightest elements: helium, hydrogen, nitrogen, and carbon dioxide. This is similar to Mercury, Venus, and Mars. After about several hundred million years, the Sun's heat and solar winds began to affect the various planets. The solar winds consist of a plasma that is composed of charged particles. These solar winds tore through the four planets closest to the Sun (Mercury, Venus, Earth, and Mars) and stripped them of their first atmosphere. The solar wind is an inverse function of the distance from the sun and loses its strength as the distance from the Sun increases. As a result, the planets beyond Mars were not stripped of their outer gases. To give you an idea of the strength of the solar winds, near Earth, the solar wind averages 270 miles a second or close to a million miles an hour.

It is important to gain a mental picture of this solar system where Earth evolved along with the other planets and the Sun. As Earth circled the Sun, like a top spinning, its axis of rotation was tilted by approximately 23.5 degrees, with respect to the plane of its orbit around the Sun. Just think how space looked 4.5 billion years ago. Think of all the objects being hurled through the universe as a result of the big bang. Imagine our cold Earth spinning at 1,000 miles an hour on its axis and orbiting around the Sun at 67,000 miles an hour. During this time of high activity in space, as a result of the big bang, many free structures were shooting through space that could, and did, impact Earth. At this point in its infancy, Earth was like an unprotected ball of material spinning on its axis and circling the Sun at a terrific speed. In the first billion years, a great amount of free material from the big bang bombarded the surface of Earth. With no atmosphere to stop this type of bombardment, plus the fact that early Earth had no magnetic field, the planet was open to this type of bombardment from free space and the Sun's solar winds.

Earth's Internal Release of Energy

As if this was not enough torture from these exterior sources, a tremendous pressure was building up inside Earth, as a result of gravity pulling the heaviest materials toward its center core. As a result, tremendous volcanoes belched out heavy materials onto Earth as well as into space. Essentially, Earth was being bombarded externally and from within. About 200 million years after the formation of Earth, an asteroid purportedly hit Earth and tore off what is now the moon.

Because we have been able to visit the moon, it is verified that the material is the same as Earth's crust and it has no core. I will discuss later how the moon has helped Mother Earth.

Now that you have this mental picture of a barren Earth with volcanoes belching hot gases and hot molten material while being bombarded externally, you wonder how it survived. We will now begin to see how this planet survived this torture. Interestingly, the torture was also the savior of Earth as we know it today.

Differentiation

The Earth began what is called differentiation.[3] This is the process whereby the heaviest materials, such as iron, gravitated to the center core, resulting in different layers being formed in the globe of Earth. It is now believed that this core contains solid iron, some nickel, and a considerable amount of gold. As the center core built up, an outer core began to surround the center core. This outer core was also composed of iron and nickel, but it differed from the hard core. It was molten iron and nickel heated by the great pressure of this compression cycle produced by gravity in segregating these layers during the differentiation process. This material of the outer core flows readily as Earth spins on its axis. Further out from this molten outer core, the next layer of differentiation began to form what is now called the mantle, which makes up about two-thirds of the volume of the planet. Several layers make up the mantle. The bottom layer is about 1,000 miles thick and comprised of heavy rocks. The penultimate layer is about 300 to 400 miles thick. Slowly moving rocks comprise this layer. Because of this movement, earthquakes occur. Counting the crust, the final layer of the mantle is about 150 miles thick. It is the layer with the tectonic plates. Under the continents, the crust is about fifty miles thick. However,

---

[3]   globalchange.umich.edu/globalchange1/current/lectures/first_billion_y...
4/28/2007

under the oceans, it is about four miles thick. All in all, it is mainly comprised of granite and basalt. This is the part that we humans and other animals live on.

Over time, due to the rough treatment that this outer layer received, it has been weathered. This included the oxidation of various rocks and minerals, storms that caused erosion, formation of glaciers, earthquakes, volcanic eruptions, tornadoes, and the bombardment by meteors and asteroids over time. The crust floats on the mantle, so it is subject to the movement of the tectonic plates. As they squeeze together, they cause mountains to be formed. When the plates are pulled apart, they create the deep basins where oceans settle. All these forces working on the crust over a few billion years resulted in what we see today as we travel the globe.

Differentiation and Earth's Magnetic Field

During differentiation, the lighter materials moved toward the surface. The mantle consisted of dense silicates and radioactive materials. These radioactive materials helped to heat Earth, and they currently continue to provide heat to Earth, but it is at a reduced rate. This differentiation resulted in a smaller size of Earth. The swirling molten iron of the outer core began to provide Earth's magnetic field. This magnetic field is the result of the spin of Earth on its axis. The swirling of this molten liquid iron around the outer core caused electric current to flow somewhat like present-day generators, resulting in the buildup of the magnetic field. This change in Earth was to be a built-in fortress to protect the planet. Solar winds could not penetrate or wipe out any new atmospheres that might be generated due to this magnetic field. The magnetic field diverts most of the solar wind. So, Mother Nature was hard at work again to take the maturing Earth to its next level. The process of this differentiation resulted in the increased pressure that forced material up toward the surface and fed the volcanoes. Eruptions from the volcanoes released gases that were to help form the second atmosphere and oceans.

The Second Atmosphere

Because of several sources, including meteorites, comets, and asteroids bombarding Earth, the rather cool planet began to heat up. Because energy cannot be created or destroyed, the kinetic energy of this bombardment was converted into heat of an equivalent amount of energy. The loss of the original atmosphere due to the solar winds resulted in the Sun's solar energy being imparted to Earth since there were no clouds to reflect the sunlight.

Another source of heat was the result of the differentiation and compression that resulted. Compression energy was then converted into heat energy.

Another source of heat was due to radioactive material in Earth's mantle and Einstein's equation that energy is equal to mass multiplied by the speed of light squared, or $E = mc^2$. As these radioactive isotopes, locked in the body of Earth, continued to radiate, they lost mass, causing the conversion of mass into energy and the resultant heat.

These radioactive isotopes continue to supply heat to Earth today, but it is at a lower level. When I say that Earth heated up, I am not talking about the temperature we enjoy today. Some estimates say that, after a couple million years, Earth's outer crust may have been molten. So we have gone through a stage where Earth was cool to a molten stage. This molten stage was the next to be transformed.

New Atmospheric Gases, Earth's Second Atmosphere

Further heat generated by differentiation continued to provide the volcanoes with heated material. They continued to spill out gases and solid material. The gases were now mostly comprised of water vapor and heavier gases consisting of gases other than helium and hydrogen, which had formed the original atmosphere. The runoff from the eruptions now consisted of Hydrogen, Helium, oxygen, nitrogen, carbon dioxide, methane, sulfur dioxide, iron oxides, silicates, water vapor, and manganese. Much of the lightweight helium escaped Earth while the hydrogen, water vapor, and oxygen began to play a different role in the formation of our present Earth. Oxygen began to oxidize the rocks of Earth's crust, thus delaying the release of the free form of oxygen required to produce an atmosphere containing oxygen. In these early stages, the oxygen was spent almost completely on the formation of oxides of the materials that comprised Earth's crust. This oxidation of the rocks was mainly on the iron present in the rocks. It resulted in oxygen being tied up in what is now called the red beds.[4]

The comets, meteorites, and asteroids were hitting Earth less frequently, but it is believed that Earth's oceans began to form from the water brought by these comets and asteroids as well as the water vapor escaping from the volcanic eruptions. The process of differentiation resulted in significant amounts of water being squeezed out of Earth during this long and involved differentiation cycle. Due to the combination of all these forces, Earth began to form another atmosphere that consisted mainly of water vapor, nitrogen, and carbon dioxide.

The water vapor from the volcanoes as well as the water brought in by space material resulted in the initial forming of the oceans. The water vapor in the newly forming atmosphere rose and hit the colder space surrounding the planet, causing a condensation of the water vapor. As a result, there were heavy rains. In fact, it was much more than what we see today. Over a significant amount of time, oceans were formed and filled the basins. The volume of water was so great that only a small amount of solid material was above the levels of the oceans. The oceans formed approximately 400 to 800 million years after the birth of Earth. The proof that there was an ocean before Earth was a billion years old is proven with the finding of single cell sea microbe fossils that are 3.5 billion years old.[5]

The forming of the oceans resulted in several phenomena. First, it cooled Earth. However, in the cooling process, Earth released heat, which converts some of the ocean's water into water vapor, which rises to the atmosphere to begin another cycle of raining. At this time in

---

[4] Geolor.com/geoteach/How_Did_Earths_Atmosphere-Evolve_geoteach
[5] Ibid.

Earth's history, there was sunlight, the atmosphere's water vapor, and carbon dioxide. All of these constituents result in global warming, which is discussed today. Here, we have the first global warming in the history of the planet. Do we have global warming? We certainly do. It has been here since Earth was about a billion years young. However, global warming at this time in Earth's life was significant due to heavier amounts of water vapor, carbon dioxide, and some methane in our second atmosphere.

The Sun's Radiation and Its Wavelengths

Global warming in Earth's early years was due to several factors. One was the lack of oxygen in the early atmosphere. The main factor that causes global warming relates to the difference in the wavelengths of the Sun's solar energy and the wavelengths that Earth radiates. The Sun's wavelengths are short. The peak energy is at about 500 nanometers.[6] The Sun's wavelengths are in the visible or near visible of our eyes, which peak at around 550 nanometers. This is one of the first steps in the evolution of man's eyesight. Because a significant amount of the Sun's output of ultraviolet is captured by the ozone in the atmosphere and infrared wavelengths are not predominant, man's eyes are not able to see these wavelengths in the form of sunshine; the infrared wavelengths, which are longer, or the ultraviolet wavelengths, which are shorter, than the 400 to 700 nanometers (0.4 to 0.7 centimeters) of our eyesight.[7]

Although this strays from my main purpose, I felt it was worth reviewing the Sun's spectrum and what we humans see while covering the subject of the wavelengths we receive from the Sun. The following, figure 1, shows the light wave's spectrum. The sunrays of light are comprised of many wavelengths. They go from 10^{-7} centimeters in wavelength to 10^{-2} centimeters.

It is important to note the wavelengths that are visible to humans. Compared to all the wavelengths the Sun emits, the sunlight is a very

---

[6] A nanometer is equivalent to one-billionth of a meter.

[7] Dev.nsta.org/ssc/moreinfo.asp?id=947

narrow spectrum of light. Note that the visible portion is brought out in the figure to show the colors that are included in these wavelengths. The red color is generated by the longest visible wavelengths just before the longer infrared wavelengths (1.0 to 10.0 microns) and violet is formed by the shortest visible wavelengths we are able to see, just before the UV (ultraviolet) light. The UV goes from 1 to 10 nanometers (billionths of a meter) just before x-rays and the gamma rays are the shortest with the most energy. Gamma rays are generated by atomic reactions.

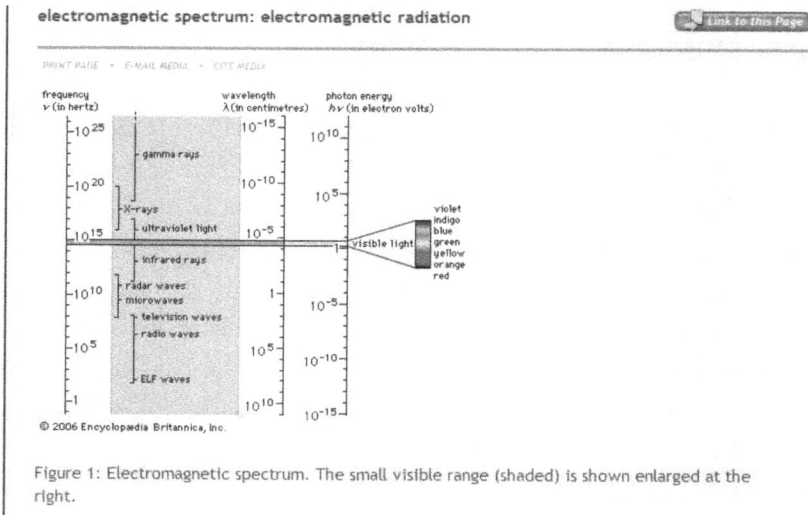

Figure 1: Electromagnetic spectrum. The small visible range (shaded) is shown enlarged at the right.

Figure 1. Electromagnetic spectrum. The small visible range (shaded) is shown enlarged at the right.

This electromagnetic spectrum represents all the wavelengths and frequencies known to man. Notice that the low frequencies have the longest wavelengths and the lowest energies for these wavelengths. The high frequencies have the shorter wavelengths and the higher energies.

The combination of all the colors is seen as white. The lack of any of these colors represents black. The energy of the short wavelengths reaches levels of 10×10^{12} electron volts (a trillion electron volts).

Spectrum of Electromagnetic Radiation				
Region	Wavelength (Angstroms)	Wavelength (centimeters)	Frequency (Hz)	Energy (eV)
Radio	$> 10^9$	> 10	$< 3 \times 10^9$	$< 10^{-5}$
Microwave	$10^9 - 10^6$	$10 - 0.01$	$3 \times 10^9 - 3 \times 10^{12}$	$10^{-5} - 0.01$
Infrared	$10^6 - 7000$	$0.01 - 7 \times 10^{-5}$	$3 \times 10^{12} - 4.3 \times 10^{14}$	$0.01 - 2$
Visible	$7000 - 4000$	$7 \times 10^{-5} - 4 \times 10^{-5}$	$4.3 \times 10^{14} - 7.5 \times 10^{14}$	$2 - 3$
Ultraviolet	$4000 - 10$	$4 \times 10^{-5} - 10^{-7}$	$7.5 \times 10^{14} - 3 \times 10^{17}$	$3 - 10^3$
X-Rays	$10 - 0.1$	$10^{-7} - 10^{-9}$	$3 \times 10^{17} - 3 \times 10^{19}$	$10^3 - 10^5$
Gamma Rays	< 0.1	$< 10^{-9}$	$> 3 \times 10^{19}$	$> 10^5$

Figure 2. Regions of the Electromagnetic Spectrum. The table above gives the approximate wavelengths, frequencies, and energies for selected regions of the electromagnetic spectrum. The notation "ev" stands for electron-volts, a common unit of energy measure in atomic physics.

This electromagnetic spectrum represents all the wavelengths and frequencies known to man. Notice that the low frequencies have the longest wavelengths and the lowest energies. The spectrum is interesting. Besides the Sun's and Earth's wavelengths, there are many of the lower and many of the higher frequencies. These wavelengths can be considered as those supplied by man or natural occurring events on Earth. Man makes sounds when he talks and sings. These sounds cover the frequencies of about five cycles per second to about 25,000 cycles per second, which are wavelengths of 1.2×10^4 meters for the high frequencies to 6×10^7 meters (37,500 miles) for the low frequencies. The low frequency wavelength is long enough to go around Earth one-and-a-half times. The spectral wavelengths that we see as colors are:

Violet	380–450 nanometers (390×10^{-9} meters)
Blue	450–495 nanometers (475×10^{-9} meters)
Green	495–570 nanometers (535×10^{-9} meters)
Yellow	570–590 nanometers (585×10^{-9} meters)
Orange	590–620 nanometers (605×10^{-9} meters)
Red	620–750 nanometers (700×10^{-9} meters)

Table 2. Spectral wavelengths seen as colors.

The colors we see are the result of the wavelengths that the object reflects. For example, when you see a green plant, it is absorbing light at all the wavelengths except the wavelengths of 570 to 590 nanometers. This wavelength is being reflected, and you are seeing that wavelength which is green. The X-rays that man re-creates use secondary emission and go from 1 x 10-8 to 1 x 10-12 meters (0.01 nanometers to 10-7 nanometers). The gamma rays are the most powerful rays. They are produced by radioactivity. An atomic explosion also releases them.

Let's return to the main subject of the short wavelengths delivered by the Sun to Earth and why this results in global warming. Notice, in addition to the short wavelengths of sunlight, a certain portion of the wavelengths is a little longer and appears as the infrared wavelengths. These wavelengths will be of interest to you in the discussion of global warming.

Earth's Radiation and Its Wavelengths

These short wavelengths of the sunlight penetrate what little atmosphere is present and encroach on the outer crust of Earth. The Earth's surface absorbs and reflects this light and its resultant heat. The oceans absorb and reflect a significant portion. At the same time, Earth radiates heat back toward the atmosphere. The wavelengths that Earth radiates are much longer than those of the Sun's radiation. This is because Earth is a cooler body than the Sun. The hotter the body of radiation, the shorter the wavelengths it radiates; and vice versa. The Earth is about twenty times cooler than the Sun's outer surface. Its radiant wavelengths are twenty times longer, or about 10,000 nanometers (10.0 microns or approximately 0.4 inches).[8]

Fortunately, this exchange of energy takes place at these different wavelengths. Plants need the short-wave to be absorbed. They do not have to worry about reradiating back toward outer space. Likewise,

---

[8] En.wikipedia.org/wiki/Photosynthesis; R.E. Blankenship, "Molecular Mechanisms of Photosynthesis," *Blackwell Science* (2002).

the radiation from Earth is in the longer waves. They do not affect the plants and their growth. This is a wonder of the physics of nature.

This reradiation by Earth of the Sun's radiation via the longer wavelengths results in selective absorption by certain molecules of gases in the atmosphere. These molecules include water vapor and carbon dioxide that resonate at frequencies generated by the long wavelengths of the infrared being reradiated by Earth below. This selective absorption of the long wavelengths results in the water vapor and carbon dioxide in the atmosphere being heated from Earth below. They, in turn, radiate their heat. Their radiation is in all directions. Thus, approximately half the heat is reradiated back to Earth. Half is lost to space. This selective absorption begins a cycle, whereby the bottom of the atmosphere is being heated. As a result, the bottom of the atmosphere is hotter than the upper atmosphere. This was the first greenhouse effect. It resulted in this atmosphere trapping the long wavelength heat and causing further warming of Earth's surface. In doing so, the oceans were heated. Additional water was vapor released to the atmosphere to continue this cycle.

It is important to understand the conditions during this time. Carbon dioxide in the atmosphere was approximately 10,000 to 100,000 parts per million, which is much denser than today's carbon dioxide levels in our atmosphere. In the meantime, water vapor, the biggest global warming gas in terms of absorption, was very high as the result of the global warming, which resulted in the oceans receiving higher heat than today through this greenhouse effect. This resulted in increased evaporation and, thus, more water vapor. It has been estimated that this combination of those gases plus methane, without the presence of oxygen, caused Earth to heat up to an average temperature of approximately 150 degrees Fahrenheit (65.5 Celsius) over a long period. This global warming of early Earth, after the heavy rains had created the oceans and cooled Earth, started to build an atmosphere that was heavy in water vapor. This set the stage for the next major metamorphous of Earth.

Photosynthesis

Fortunately, Earth contained several things required to establish a major change. Either as part of its original makeup or deposited by other space debris; it had RNA (ribonucleic acid) and DNA (deoxyribonucleic acid); plus the equivalent of seeds that would produce the vegetation that became the next step in Earth's evolution. With the combination of the sunlight, water vapor, and carbon dioxide, the stage was set for the process of photosynthesis to occur.

Simply, photosynthesis is the conversion of light energy into chemical energy. Photosynthesis occurs in plants in two stages. In the first phase, light-dependent reactions capture the energy of light and use it to make high-energy molecules. During the second phase, the light-independent reactions use the high-energy molecules to capture carbon dioxide and make the precursors of glucose.[9]

In the light-dependent reactions, one molecule of the pigment chlorophyll absorbs one photon of light and loses one electron The chlorophyll molecule regains the lost electron by taking one from a water molecule through a process called photoysis, which releases oxygen gas as a waste product. During this process ATP is synthesized. ATP (Adeninetriphosphate) is the fuel in the human body or in plants that provides energy for muscle contraction and to move energy into and out of cells.

The Earth received a miracle. The process of photosynthesis begins to eliminate the atmosphere of its carbon dioxide, which was causing the greenhouse effect, while producing free oxygen to begin a change in the atmosphere's contents. Plants begin to grow in the solid Earth and sea kelp begins to grow in the oceans. Algae come in multiple forms from multicellular organisms, like kelp, to microscopic, single-celled organisms. Photosynthetic bacteria do not have chloroplasts. Instead, photosynthesis takes place directly within the cell.

The ability to convert light energy to chemical energy confers a significant evolutionary advantage to living organisms. Early

[9] Space.com/searchforlife/life_origins_001205.html

photosynthetic systems, such as those from green and purple sulfur and green and purple nonsulfur bacteria, are thought to have been anoxygenic (during photosynthesis they do not produce oxygen, as such they take in carbon dioxide but do not give off oxygen), using various molecules as electron donors. Green and purple sulfur bacteria are thought to have used hydrogen and sulfur as an electron donor. Green nonsulfur bacteria used various amino and other organic acids. Purple nonsulfur bacteria used a variety of nonspecific organic molecules. The use of these molecules is consistent with the geological evidence that the atmosphere was highly reduced at that time. Fossils have been found of what are thought to be filamentous photosynthetic organisms dating from 3.4 billion years ago.

It is important to realize the other advantages of this phase of Earth's existence. Most plants are photoautotrophic, that is, they can synthesize food directly from inorganic compounds using light energy, for example, from the Sun, instead of eating other organisms or relying on nutrients derived from them. The most important contributors of free oxygen into the Earth's atmosphere were cyanobacteria (blue-green algae). The oldest known fossils were found in rocks of Western Australia and were dated 3.5 billion years old. These were also photoautotrophic. Through photosynthesis, these bacteria convert carbon dioxide directly to oxygen through a chemical reaction. Another important aspect is the fact that plants take in nitrogen through their roots. Thus, the nitrogen becomes part of the plant's system. Human beings, even though they were not on Earth yet, need nitrogen, but they are unable to take in nitrogen from the air. We breathe it in, but it is expelled without being assimilated. To receive our needed nitrogen we eat plants or other animals that eat plants. So, photosynthesis brought many good aspects to the aging Earth in its early existence, and it still does.

While photosynthetic life reduced the carbon dioxide content of the atmosphere, it also started to produce significant amounts of oxygen. For a long period, the oxygen produced did not build up in the atmosphere. The oxygen oxidized rocks on the Earth's surface continuously for a significant amount of time. As recorded in banded iron formations and continental red bed formations, the rocks took it up. Only about a billion years ago did the reservoirs of oxidizable rock became saturated and the free oxygen stayed in the air. By this time, forests, oceans, and jungles throughout Earth were supplying large amounts of oxygen. Thus, the atmosphere evolved toward what is our present atmosphere.

The Third Atmosphere

So, Mother Earth has accomplished several tricks, starting with differentiation to provide Earth with a magnetic field and generate heat that resulted in volcanoes belching out water vapor and carbon dioxide that supplemented the building up of the oceans and the second atmosphere. This atmosphere reacted with the sunlight to provide photosynthesis, which resulted in the beginning of the growth of plant life in the oceans and, eventually, on solid ground. While the oceanic plant life began the reduction of the carbon dioxide in the atmosphere, it also began to supply oxygen for the beginning of a new third atmosphere, one devoid of most of the carbon dioxide. With this last trick came the eventual elimination of the strong greenhouse effect. Well, not quite. We will discuss the remaining greenhouse effect later.

Meanwhile, this was the beginning of the growth of plant life on Earth, both on land and in the seas. The ocean received organic matter from the land and the atmosphere as well as from incoming meteorites and comets.

The combination of substances received began forming significant molecules, such as sugars, amino acids, and nucleotides, the building blocks of proteins and nucleic acids. Somewhere along the line, DNA and RNA entered into the scheme of things.[10] DNA provides the instructions for the building blocks of nature's living organisms through the use of the RNA messenger. It started the chemistry working toward a future goal. It also gave us another surprise, single-celled microorganisms. Single-celled microbes have been identified in rocks that date back to

---

[10] En.wikipedia.org/widi/Cambrian; Stephen Jay Gould, *Wonderful Life: the Burgess Shale and the Nature of Life* (New York: Norton, 1989).

3.5 billion years, just a billion years after Earth's existence. To date, we have only been able to identify multicellular organisms having existed a billion years ago.

The Earth, through photosynthesis, continued to reduce carbon dioxide and replace the atmosphere with oxygen. With the oxygen and nitrogen taking over a major role in the atmosphere, ultraviolet light from the Sun split the oxygen molecules in the atmosphere and produced an ozone layer in the upper part of the atmosphere. As this continued, an ozone ultraviolet shield was produced as a by-product. This shield prevented most of the harmful ultraviolet from reaching Earth. This occurred about a billion years ago and set the stage for oxygen-breathing life forms. Any life form of animals before the creation of the ozone layer would have found most, if not all, of that life form dying due to the ultraviolet rays from the sun. It has been recognized that living things as we know them can not take the ultraviolet rays and their extreme energy. The equivalent of skin cancer or other forms of cancer would have caused the death of any living organism before ozone was produced in the atmosphere.

The Cambrian Explosion

Over the next few billion years, Earth evolved toward what we now experience. However, there were no creatures with bones or shells. We have never found any evidence through fossils or any other means that anything existed, except for multicellular organisms, until about 540 million years ago. They came about in a time that we call the Cambrian Explosion.[11]

However, well before the Cambrian Explosion, Earth's landmass began to change shape as large, stable continents took on the smaller land areas and their volcanoes. The large landmasses were more stable because they represented interiors that were less affected by the movement of the tectonic plates and were somewhat free of volcanic action. This is

---

[11] En.wikipedia.org/wiki/Oxygen; Neil A. Campbell and Jane B. Reece, *Biology*, 7th ed. (San Francisco: Stanford University Press, 2005).

true today when you review where most of earthquakes occur as well as volcanic action. These actions occur on the edges of continents, mainly the ones that interface with the Pacific Ocean. As the surface continually reshaped itself over hundreds of millions of years, continents formed, broke up, and formed super continents. It has been hypothesized that severe glacial action between 750 and 580 million years ago covered much of the planet in a sheet of ice. This hypothesis has been termed "snowball earth." It is of particular interest because it preceded the Cambrian Explosion when multicellular life forms began to proliferate.

The Cambrian Explosion occurred about 535 million years ago, as indicated by fossils that showed, for the first time, that marine invertebrate, such as shell-making ammonites, appeared. Fish, amphibians, reptiles, birds, and, eventually, mammals followed. The Cambrian Explosion is the earliest period where rocks are found with numerous large, distinctly fossillizable, multicellular organisms that are more complex than sponges or medusoids. During this time, roughly fifty separate major groups of organisms or phyla[12] emerged suddenly. In most cases, it was without evident precursors.

Though this life explosion began in the Earth's seas, it can be attributed, if only in part, to the increased free oxygen levels in the warm, stable marine and atmospheric environments. As mentioned, once oxygen existed in the atmosphere, the Sun's ultraviolet rays split the oxygen into ozone. The ozone shield was created. At this point, life moved out of the oceans and evolved into land-dwelling organisms that were capable of respiration. Likewise, the land plants reached another level of sophistication with plants like conifers and flowering plants. It is not known why this step-like function of complex living things occurred in the Cambian Explosion.[13] This time period still remains somewhat a mystery to be solved. (The Cambrian is named for Cambria, the classical name for Wales, the area where rocks from this time period were first studied.)

---

[12] A phylum defines the basic body plan of some group of modern or extinct animals.
[13] The Cambrian is named for Cambria, the classical name for Wales, the area where rocks from this time period were first studied.

I would make an educated guess at what happened during the Cambrian Explosion. At some point, the amount of atmospheric oxygen that evolved as a result of photosynthesis had an effect on the oceans and the oxygen that made up the water at that time. It allowed the oceans to reach a form of balance of the oxygen in the water and the oxygen in the atmosphere. Perhaps it affected the amount of water vapor in the atmosphere. This would have made a transition from the oceans to the atmosphere through a media that was similar to the ocean. We are not talking about very sophisticated life forms as we recognize today. Compared to what had gone on before, it was complex. Evolution began as more oxygen was available and animals grew more complex and fought for their position in the food chain. On the other hand, this transition may relate to the atmospheric pressure obtained in this balance since atmospheric pressure is just the weight of the atmosphere above the ocean level. As more oxygen was generated into the atmosphere, it must have increased the weight of the atmosphere at the sea level which is recognized as atmospheric pressure. There may have been a shift to a higher percentage of heavy water at that time, which is $h_2 0_2$. Or it may have resulted in the water content having less heavy water. Heavy water is found at present in water in about one part in a thousand to one part in ten thousand molecules of standard water at different locations on Earth. The fact that oxygen existed in the water for some time and then the atmosphere began to achieve some level of oxygen and water vapor resulted in a smooth transition from the ocean to outside air.

It is interesting to consider that Mother Nature was up to some more tricks. With an oxygen atmosphere having evolved from the output of plant life in the oceans, there was less and less carbon dioxide available for the plants. Plants cannot live without at least 50 parts per million of carbon dioxide in the atmosphere. Maybe this was Mother Nature's way of providing animals that would breathe out carbon dioxide to provide a balance. With the arrival of animals that breathed carbon dioxide it, began to provide the carbon dioxide needed for the plants to exist.

The Buildup of Oxygen in the Atmosphere

Before the evolution of water oxidation in photosynthetic bacteria, oxygen was almost nonexistent in Earth's atmosphere. Free oxygen first appeared in significant quantities between 2.5 and 1.6 billion years ago as the product of the metabolic action of early anaerobes, for example, archaea and bacteria. These organisms developed in the mechanism of oxygen evolution between 3.5 and 2.7 billion years ago. At first, the produced oxygen dissolved in the oceans and reacted with iron. It started to gas out the oxygen-saturated waters about 2.7 billion years ago, as evident in the rusting of iron-rich terrestrial rocks that started about that time. The amount of oxygen in the atmosphere increased gradually at first and shot up rapidly around 2.2 to 1.7 billion years ago to about 10 percent of its present level, which was still quite a low oxygen level of less than 2 percent.

The development of an oxygen-rich atmosphere was one of the most important events in the history of life on Earth. During the oxygen catastrophe about 2.4 billion years ago, the presence of large amounts of dissolved and free oxygen in the oceans and atmosphere may have driven most of the anaerobic organisms (organisms that can not breathe oxygen) that were then living to extinction. However, the high electro negativity of oxygen creates a large potential energy drop for cellular respiration, thus enabling organisms using aerobic respiration to produce much more ATP[14] than organisms. This makes them so efficient that they have come to dominate Earth's biosphere. Photosynthesis and cellular respiration of oxygen allowed for the evolution of eukaryotic cells and, ultimately, complex multicellular organisms like plants and animals.

If we look at the last 4.5 billion years of Earth's existence, we see an atmosphere with less than three parts per million of oxygen; four billion years ago, this rose to approximately 9,000 parts per million. Three billion years ago, it continued a rise to approximately 900,000 parts per million when oxygenic photosynthesis began to raise the level in the atmosphere. It stayed on this kind of linear growth curve to

---

[14] ATP is considered to be the fuel for human and animal growth.

approximately 1.5 billion years ago. At that time, there was a dramatic shift in the rate, which showed approximately 1.5 percent, or 1.5 million parts per million, oxygen in the atmosphere. Different types of plants grew. Approximately 540 million years ago during the Cambrian era, the amount rose to close to 10 percent. Shortly after that time, there was a slower growth to the more than 20 percent of today.[15]

During this slow growth period, animals began to appear on Earth. I assume the slower growth in the atmospheric oxygen was due to this appearance of animals, which began the respiration of oxygen and generation of carbon dioxide.

The atmospheric abundance of free oxygen in later geological epochs and its gradual increase up to the present has been largely due to synthesis by photosynthetic organisms. Over the past 500 million years, oxygen levels fluctuated between 15 and 35 percent per volume.

Toward the end of the Carboniferous era, that is, coal age, about 300 million years ago, atmospheric oxygen levels reached a maximum of 35 percent by volume, allowing insects and amphibians with limited respiratory systems to grow much larger than today's species. Today, oxygen is the second most common component of Earth atmosphere[16] after nitrogen. Algae and green microorganisms in the oceans produce about 75 percent of the free element that are being produced by algae and green microorganisms in the oceans. The other 25 percent is from terrestrial plants.

The following graph shows the buildup of oxygen over the past 550 million years and lists the disposition of the oxygen.[17] Molecular oxygen is essential for cellular respiration in all aerobic organisms. It is used as electron acceptor in the mitochondria to generate chemical energy in the form of ATP, a multifunctional nucleotide that is most important as an intercellular energy transfer. It is produced as an energy source during the process of photosynthesis during oxidative phosphorylation.

---

[15] en.wikipedia.org/wiki/Cambrian; Stephen Jay Gould, *Wonderful Life: the Burgess Shale and the Nature of Life* (New York: Norton, 1989).

[16] About 21 percent of volume

[17] En.wikipedia.org/wiki/Oxygen; Neil A. Campbell and Jane B. Reece, *Biology*, 7th ed. (San Francisco: Stanford University Press, 2005).

During this reaction, oxygen is reduced to water. Conversely, free oxygen is produced in the biosphere through photolysis[18] of water during photosynthesis in cyanobacteria, green algae and plants, thus closing the biological water-oxygen redox cycle.

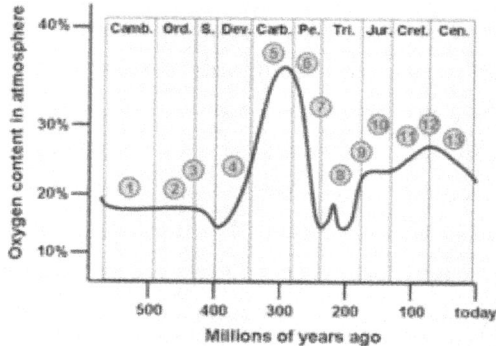

Figure 3. Oxygen content in atmosphere.

Fluctuations of oxygen levels in the atmosphere over the past 500 million years, with accompanying events, include:

- Radiation of animal phyla (Cambrian Explosion)
- First land plants
- Ordovician-Silurian extinction events
- Huge forests form on land; first land animals and seed plants
- Coal formation; first conifers, insect and amphibian gigantisms (larger than any time before)
- Low ocean levels; Pangaea forms
- Permian-Triassic extinction event
- First primitive flowering plants and dinosaurs
- Triassic-Jurassic extinction event
- Age of dinosaurs
- Radiation of flowering plants
- Cretaceous-Tertiary extinction event
- Radiation of mammals

---

[18] Light-driven oxidation and splitting

With these types of phenomena occurring, sea life began its evolutionary process. Its life form evolved into more complex life forms. Later, fossil findings show that the sea life gained the ability to breathe oxygen. There is an evolutionary metamorphic change, whereby amphibians began to leave their water homes and migrate onto land. Thus, this began the major change of life form on Earth whereby large, oxygen-breathing animals began to roam Earth. Most fossil findings show there were small and large oxygen-breathing life forms that evolved after the Cambrian Explosion.

Evidence shows there have been several mass extinctions since the Cambrian Explosion. The last occurred 65 million years ago when a meteorite collision probably triggered the extinction of the dinosaurs and other large reptiles. But it spared small animals, such as mammals, which then resembled shrews.[19]

This meteorite collision may have eliminated the large reptiles, but it had to result in Earth gaining added energy. This resulted in a huge gain of energy as a result of Einstein's equation ($E = mc^2$). The large mass of this meteorite plus the kinetic energy was converted into heat and potential energy. This added energy eventually resulted in several phenomena occurring. One was the transformation of the dead carcasses into oil and the energy that the oil supplied. The second effect was to fill the atmosphere with dust, which helped to cover the carcasses as they fell back to Earth. This dust cut off the sunlight and oxygen from Earth and caused the death of the large reptiles.

Several millions year ago, an African ape-like animal gained the ability to stand upright. It is hypothesized that this allowed for the use of its free limbs to perform tasks that were more complex. This further improved on its learning curve. Its brain became more advanced. From this step in evolution, there are several theories as to how its brain increased in size for its small body size. There was a major step in the

---

[19] En.wikipedia.org/wiki/Oxygen; Neil A. Campbell and Jane B. Reece, *Biology*, 7th ed. (San Francisco: Stanford University Press, 2005); G.B. Dalrymple, *The Age of the Earth* (San Francisco: Stanford University Press, 1991).

development of agriculture and then civilization, allowing humans to influence the Earth in a short time span as no other life form had.

The World without People or Oxygen-Breathing Animals

So far, I have provided the basics that show how the world works. This discussion has been rather straightforward because it is comprised of history and how the world got to where it is and how it coped with certain problems. At least it's the scientific view or hypothesis that is taught today. Essentially, it is how Mother Nature handled the problems without man's interference.

In order to present the basics of Earth, I have simplified the picture until now. In order to better understand the many changes that have happened in the recent history of Earth and try to understand the fantastic capability of nature, I am going to go back just a little in time to discuss the power of this planet. I want you to realize, without man being involved, there have been tremendous happenings that are related to the climate. I am going to go back to the time after the meteor struck this world 65 million years ago.

Climate from 65 Million Years Ago

From 65 to 24 million years ago, it was quite warm. Measurements made on oxygen isotope ratios show the ocean to have been much warmer. In the Atlantic Ocean, it was about 80 degrees Fahrenheit (26.6 degrees Celsius). In the Artic Ocean, the temperature was 73 degrees Fahrenheit (23 degrees Celsius). Can you imagine going to the Artic Ocean and seeing things growing like you now see in Florida? Canada had an average temperature of 65 degrees Fahrenheit (18 degrees Celsius).

This type of climate resulted in a complete elimination of some species of animals and sea life, which was similar to the extinction of the dinosaurs. Scientists do not know what caused this dramatic effect on our environment and climate. Each scientist has a different hypothesis of what could have caused this dramatic shift in climate. I would take an educated guess that, when the meteor struck Earth and killed the dinosaurs, it filled the atmosphere with a great amount of dust, which collected water vapor. A long period of global warming occurred as water vapor in the atmosphere absorbed the long waves of light around the infrared and near infrared and reradiated it back to Earth.

Remember photosynthesis and how it provided Earth with oxygen? During this time, no animals or people were on Earth. So, the plants that grew were putting out oxygen and consuming the large amount of carbon dioxide in the atmosphere. This occurred over several billion years and brought the carbon dioxide level down tremendously. This continued until the Cambrian Explosion. About a few million years after this, approximately 300 million years ago, animal life and, eventually, man began to appear on Earth. This may have been another of Mother Nature's tricks to balance Earth. This is a guess on my part, but plants cannot live if the carbon dioxide level drops below 200 parts per million for any length of time. Some plants can live down to 50 parts per million. It is possible that the plants had stripped the large amounts of carbon dioxide that had been in the atmosphere. The atmosphere was close to 200 parts per million or lower. Then animal life came along. Eventually, man began generating the carbon dioxide level back to an increased level. If this scenario is true, then Earth owes its life to the fact that oxygen-breathing animals appeared. Since that time, a balance of animals and plants provide this carbon dioxide level to remain in the comfortable level of 250 to 800 parts per million and Earth to gain a larger oxygen atmosphere to satisfy the animal's needs.

I have not read any reports that mention this information. It is just a current observation of mine. It makes sense. If this is the case, when the meteor struck Earth 65 million years ago and destroyed the large animals, Earth warmed up because of the breaking of this balance. This warming might have been due to the dust in the air and the increased water vapor that this dust would have trapped in the atmosphere. A large increase in water vapor will cause global warming faster than an increase in the carbon dioxide level.

The temperature started to drop around 34 million years ago. This drop continued for 10 million years. Glaciers began forming in what was then Antarctica. The growth of ice sheets led to a drop in the sea level. Areas that had been tropical for many years became much colder. Forests were initiated in many parts of Earth that had been covered with tropical plants. Some blame this on tectonic movement because this is when India collided with Asia and pushed the Himalayas to their height as the highest mountains in the world. This shoving of the Himalayas to the great heights kept the moist air from reaching parts of the world. It is hypothesized that this caused the Gobi desert and one other large desert to be formed.

At this time in Earth's history, a huge continent comprised all of what are now Australia, South America, and Antarctica. During this time period, this huge continent broke off into three continents: Australia, South America, and Antarctica. As a result, water completely surrounded Antarctica. Some scientists believe this resulted in the colder climate because of this change in the flow of water from the area around Antarctica.

I am going to skip some of the major events that occurred between 24 and 1.8 million years ago when the Earth entered an even cooler phase, which resulted in the temperatures moving up and down over periods of about 100,000 years. In the last million years of this period, there were ten excursions between ice and the lack of ice. These were rather short ice ages. Experts call them glacials. The intervening periods are called interglacials. I am really talking about 100,000 years between excursions, but this is rather abrupt when you consider how long it took for other major climate changes since Earth's birth.

The latest glacial began around 70,000 years ago. People were around then. The Neanderthal died out around 35,000 years ago. The oldest cave paintings are 32,000 years old. The glaciers reached their maximum extent around 21,000 years ago. Ice sheets extended down to the Great Lakes and the mouth of the Rhine. They also covered the British Isles.

Around 16,000 years ago, temperatures started to seriously warm up. They reached their present levels around 10,000 years ago. The temperature is now becoming hotter than it has been since the end of the last glacial. With just 2 more degrees Fahrenheit (about 1 degree

Celsius), we will have reached a global temperature that is hotter than any time in the last million years.

Even though the Earth has been warming up in the last 16,000 years, there was a cold spell between 12,900 and 11,500 years ago. This drop in temperature lasted for 1,400 years, during which time the temperatures dropped dramatically in Europe. It was about 12.6 degrees Fahrenheit (7 degrees Celsius) in only twenty years. In Greenland, it was 27 degrees Fahrenheit (15 degrees Celsius) colder than today. In England, the average annual temperature was 23 degrees Fahrenheit (-5 degrees Celsius). Glaciers started forming. Scientists provide several different scenarios as to what caused this change. One theory is that something blocked a current of water called the Atlantic thermohaline circulation, which brings warm water up the coast of Europe. Proponents of this theory argue that this current makes London much warmer than Winnipeg, Canada, or Irkutsk, Russia, in today's climate. Some of today's scientists are concerned that we could see a repeat if the Arctic Ocean continues to lose ice. Some scientists believe the Atlantic Ocean is not the Achilles heel of the climate system. There are other reasons, such as the winds that move over this area of Europe.

From 6,000 to 2,500 years ago, things cooled down with the coolest stretch occurring between 4,000 and 2,500 years ago (only 500 BC). Things warmed up for a while. Then they cooled down from 500 to 1300 (the Dark Ages).

During a period from 1000 to 1300, European temperatures reached their warmest levels for the last 40,000 years. From 1450 to 1890, there was a cooling period. This is often called the "little ice age." This killed off the Icelandic colonies in Greenland, as described in *The Greenlanders* by Jane Smiley. However, this might have been a more localized situation with only a slight drop in Europe of less than 1 degree Celsius (about 2 degrees Fahrenheit).

Since then, the climate has been warming. Most scientists believe this warming is much greater than the other variations over the last 1,000 years. In general, it is the consensus of opinions that the Northern Hemisphere has shown a rise in temperature by about 2 degrees Fahrenheit (1.1 degrees Celsius) from 1900 to 2000. 1998 was the hottest year.[20] 2005 is close behind. 2006 did not show this trend. There were much fewer hurricanes in 2006 than the previous few years. Sometimes, based on the damage they do, hurricanes are overstated as to their strength. For example, the fantastic damage done by Hurricane Katrina in 2005 was not the result of the strength of the hurricane. More accurately, it was the unique situation that was relative to New Orleans. This area, which is below sea level, and the path of the hurricane brought water levels over the levees. As a result, the most damaging hurricane to strike the United States occurred. Many years before this, there were recommendations to restructure this area to prevent the kind of damage done. The Netherlands, or Holland, is below sea level and major construction in that country has overcome most of the danger there.[21]

This chronology about the major climate events over the last 65 million years provides you with information about how Earth has bounced around on climate and temperatures without man's influence. I wanted you to understand the kind of major changes that have occurred on Earth without any influence by man so I could provide information on what nature brings to the present picture of climates around the world. There were other sections where the energy of Earth, Sun, and weather brought their powers to be with dramatic results. With this type of background information as a reference, I will now bring man into the picture. Before this, I want to provide some information that will eventually impact man and his stay on this planet. To do this, I must describe entropy and its major eventual affect on Earth and mankind.

Entropy

The basic law of physics is that energy cannot be created or destroyed in a closed system. This has proven true over the years. The one phenomenon that is left out of this equation is that energy can be unusable within

[20] There was an El Niño that year.

[21] Much of the information in the section was provided via an Internet article titled "Temperature" by John Baez.

that system. It is still around, but it is in an unusable state. It is not created or destroyed. It is simply unused. When there is a transfer of energy to another form, it does so evidently with the loss of energy in the total transaction. But the energy is not lost or destroyed. It is just unusable in the new state. To describe this in simple terms, I will give a simple example. Let's assume a large boulder of 5,000 kilograms is sitting on a high hill about 1,000 meters in height. This large boulder has a very high potential energy. The force of gravity that is exerted on an object times its height above sea level determines it potential energy. The pull from the center of Earth on the given object determines the force of gravity.

So, this large boulder is sitting on the high hill. It has a potential energy determined by its distance from the center of Earth. However, because the center of Earth is common for any object on Earth, potential energy is determined by its height above the surface of Earth. The equation for potential energy is energy equals mass (in kilograms) multiplied by the pull of gravity (9.8 meters per second squared) multiplied by the height above the surface of Earth (in meters), therefore, PE = h x m x g. This energy is measured in joules. When an object goes from one height (h1) to a lower height (h2), it loses some potential energy, which the difference in height (in meters) determines. If the object retains its mass and the gravitational pull is the same in its new location, then the new potential energy is just the difference in height. Therefore, PE = $(h_1 - h_2)$ x (mass x g). The difference in height is 1000 meters. The mass is 5,000 kilograms. The pull of gravity is 9.8 meters per second squared.

Let's say that some force acts on the boulder. It rolls down the hill to a position at 20 meters above sea level. It has lost a significant amount of its potential energy by going from 1,000 meters to 20 meters (relative to sea level). The energy lost is 980 meters multiplied by 9.8 meters per Second Square multiplied by 5,000 kilograms. This energy difference is expressed in joules (or watts which are joules per second). However, the boulder sits there. It still has a large potential energy. However, since it is at the lowest point in the general vicinity, it cannot roll to a lower height. Therefore, it sits there with unusable energy. It has gained in entropy. Whenever energy is lost in a system, it gains in entropy if the remaining energy is in an unusable state after the transfer. It may have smashed some houses or whatever in its path down the hill. Therefore, it transferred some of its potential and kinetic energy to the destruction of these houses. But it still sits there at sea level with unusable energy. Entropy increase occurs many times in the transfer of energy. If it is every recovered, it must be noted it is from entropy and the entropy has decreased.

Now let's look at a real situation. Remember the large meteor that struck the Earth 65 million years ago and killed the dinosaurs. The loss of the dinosaurs represented a large increase in entropy because they lost the energy of life and now obtained energy that was unusable as it lay lying beneath the dust that this large impact created. Most of the energy that Earth gained was due to this heavy object carrying a terrific amount of kinetic energy.[22] In this case, the meteor had a significant amount of mass (m). It was moving at a very high velocity. This energy was transferred to Earth in the form of kinetic energy (KE = ½ mv²), resulting in the earth gaining energy along with heat and mass. The increase in entropy afterwards was due to the dead carcasses lying under the dust that collected on their bodies for years. This high entropy state lasted approximately 65,001,850 years before it started to be useable energy (in the year 1859 the first commercial oil well was drilled). This was when it started to go from entropy to usable energy. As the energy is consumed the entropy decreases until it becomes negligible. The manner in which this energy was used will be discussed later. You will find entropy interesting on how it impacts global warming and another more difficult problem.

We have seen how Earth went through many changes and how the third atmosphere was formed. While proceeding through these various changes, there were major shifts in the climate. In this review, there were no short-term shifts in the climate that could explain what is presently being forecasted. Most climate shifts were dramatic shifts that took millions of years to complete. They resulted in, or were a part of,

---

[22] Kinetic energy equals mass multiplied by velocity squared and divided by two.

major changes in Earth. Because of the nature of reviewing history that long ago, it is difficult to ascertain or reveal small changes in climate or evolution. Only major changes can be seen. Therefore, there is no evidence one way or another, to show us what may be portrayed as the beginning of a shift. We can only see what has happened after the shift had occurred. Meanwhile, we find ourselves in essentially the same position today, where it is difficult to determine if today's climate and any change is the beginning of a long-term, gradual, or abrupt shift.

Based on a better understanding of our world and what impacts it, we can hypothesize. We can model what we know and test the model. Based on the models, we can make predictions. However, models evolve by finding either errors in the model or unknowns that should have been in the model. So, the modeling is not a one-shot episode. It is a learning curve that allows us to improve on it and approach what we believe may be an adequate one. From that, we work on improving the model as we gain more information.

There was the time during the second atmosphere that we had significant global warming as a result of heavy amounts of carbon dioxide and water vapor in the atmosphere. It does prove that the constituents of the atmosphere can cause global warming. It verifies excessive amounts of carbon dioxide and/or water vapor in the atmosphere can cause global warming. After reviewing this background of Earth's history and seeing how Mother Nature has coped with major issues in the various stages of Earth's evolution, we can now analyze the recent past where Earth has essentially settled down and take a look at the present state of Earth and how man may be affecting it.

Understanding the Stabilization of Earth as Man Enters the Scenario

Before determining what man has brought to the evolution of Earth, it is important to review the recent history of Earth and the various elements of its composition. Earth has reached a level of stabilization over the last few million years. It is worth detailing this stabilization as it provides Earth with a certain amount of built-in inertia that must be overcome before there are any changes in the climate that man's activities might initiate. I will review Earth's present atmosphere, Sun/Earth energy budget, main contributors to today's weather, mass budget, and all the phenomena that relate to Earth as we know it today. This is Earth as man began his journey.

The present pattern of ice ages began about 40 million years ago. Then it intensified about 3 million years ago. The Polar Regions have since undergone repeated cycles of glaciation and thaw, repeating every 40,000 to 100,000 years. The last ice age ended 10,000 years ago. The ice never reached areas such as Africa. It is believed this gave a head start on the evolution of human beings in Africa. The history of man has been fairly well-documented since that time. The basic body plans of the major animal phyla are established over a relatively short period of roughly 10 million years. All the major animal phyla that exist today, about thirty-six, evolve from the Cambrian faunas.[23]

The Sun, Earth's Constant Companion

Throughout time, the Sun has been Earth's constant companion and almost the sole source of Earth's energy. The only other sources of energy came many years ago when space material collided with Earth. Even today, a relatively small amount of space material is being added to Earth as very small meteorites burn up as they enter Earth's atmosphere. Radioactive material within our mantle provides energy to our world, but it has reduced significantly over time. We can discount this small amount of energy and consider our Sun as our sole source of energy. It is a constant source of energy as it continues to manufacture energy via the atomic reaction that Einstein's equation described.

The energy is generated in the Sun's core as the terrific pressure due to its internal gravitational pull causes its internal core to be extremely hot. This heat activates the hydrogen and fuses four hydrogen atoms into

---

[23] Animals or animal life especially of a period or environment

helium atoms through nuclear fusion on a continuous basis.[24] Because the atomic mass of the helium is less than the combined mass of four hydrogen atoms, the mass loss provides enormous amounts of energy via Einstein's equation. This energy release must make its way to the Sun's surface via countless absorptions and reemissions. The distance it must travel to reach the surface is enormous due to the Sun's size and the fact it makes this journey via a random route. It is estimated that it takes between 55 and 65 million years to make it to the surface. So the sunshine we see today was released when the dinosaurs were roaming Earth approximately 65 million years ago. So, even if the Sun stopped generating the power internally, Earth would receive the power already generated for approximately 60 million years. It isn't about to stop now. The Sun is the heat engine that drives the circulation of our atmosphere. Although it has long been assumed to be a constant source of energy, recent measurements of this solar constant have shown that the Sun's base output can vary by up to 0.02 percent over the eleven-year solar cycle. Temporary decreases of up to 1.5 percent have been observed. (54) Atmospheric scientists say this variation is significant and it can modify climate over time. Plant growth has been shown to vary over the eleven-year sunspot and twenty-two year magnetic cycles of the sun, as evidenced in records of tree rings.

While the solar cycle has been nearly regular during the last 300 years, there was a period of seventy years during the seventeenth and eighteenth centuries when very few sunspots were seen, even though telescopes were widely used during this period. This drop in sunspot number coincided with the timing of the little ice age in Europe, implying a sun-to-climate connection. Recently, a more direct link between climate and solar viability has been speculated.[25] I am reviewing these kinds of actions to determine, if possible, if the Sun went through some recent exercise that could result in global warming.

---

[24] en.wikipedia.org/wiki/Solar_radiation; www.grida.no/climate/ipcc_tar/wg1/041. htm#121; "Construction of a Composite Total Solar Irradiance (TSI) Time Series from 1978 to present." This article incorporates text from the Encyclopedia Britannica Eleventh Edition article, "Sun," a publication now in the public domain.

[25] okfirst.ocs.ou.edu/train/meteorology/EnergyBudget2.html

Stratospheric winds near the equator blow in different directions, depending on the time in the solar cycle. Studies are underway to determine how this wind reversal impacts global circulation patterns and weather.

Solar radiation is a radiant energy that the Sun emits. It emits a spectrum of solar radiation close to that of a black body with a temperature of 5800 degrees Kelvin. Normal temperatures on Earth are between 273 degrees Kelvin[26] and 373 degrees Kelvin.[27] Therefore, there is about a twenty-to-one difference. It relates to the difference in radiation from the Sun versus the radiation of Earth. About half of the Sun's radiation is in the visible shortwave part of the electromagnetic spectrum. The other half is mostly in the near infrared part with some in the ultraviolet part of the spectrum. The ozone in our atmosphere reflects and absorbed most of the ultraviolet. The ultraviolet radiation not absorbed by the atmosphere or other protective coating is responsible for the change of color in skin pigments.[28] It is the major cause of skin cancer.

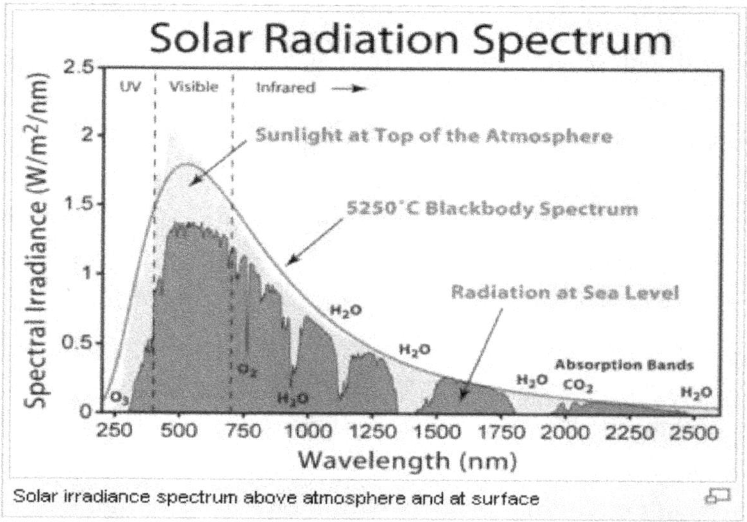

Figure 4. Solar irradiance spectrum above atmosphere and at surface.

---

[26] The freezing point of water

[27] The boiling point of water

[28] okfirst.ocs.ou.edu/train/meteorology/EnergyBudget2.html

Note that the visible sunshine is from 400 to 700 nanometers in wavelength. The short wavelengths are marked with O_3. This represents the ozone that absorbs the short wavelengths of the ultraviolet. The light area is the total sunlight at the top of the atmosphere. The dark level is what Earth receives. Note that the infrared are the wavelengths that go from just below the 1000 nanometer level to the end of the chart. They are greatly reduced in irradiation from the energy of the visible, but there are enough to not be ignored.

Solar Constant, the Sun's Energy to Earth

The solar constant, the amount of incoming solar electromagnetic radiation per unit area, is measured on the outer surface of Earth's atmosphere in a plane that is perpendicular to the rays. It is called the solar constant because it provides this amount of energy constantly. The solar constant includes all types of solar radiation, not just the visible light. It covers the previous spectrum shown earlier. Via satellite, it is measured to be roughly 1,366 watts per square meter.[29] It fluctuates by about 6.9 percent during a year, that is, it is 1,412 watts per square meter in early January and 1,321 watts per square meter in early July. It changes by a few parts per thousand from day to day. Thus, for the whole Earth, with a cross section of 127,400,000 square kilometers, the power is 1.740×10^{17} watts, plus or minus 3.5 percent. This is the amount of energy given to Earth as seen from the Sun. The Sun only sees about one- fourth of Earth at any given time. (The Sun's light is radiant, meaning it is direct line of sight. If one were to take a picture of any sphere, he would only see one fourth of the sphere. This is a fact of geometry.) Fortunately, Earth is spinning. During one cycle, that is, one day, this amount of energy is exposed to the whole Earth. Because of the angle of Earth, there is a different exposure for different parts of Earth during all of the four seasons as earth spins on its orbit around the sun.

The solar constant is relatively constant, but it varies according to

---

[29] physicalgeography.net/fundamentals/7j.html

sunspot activity. It mainly affects long-term climates rather than short-term weather. The Earth receives a total amount of radiation, which is determined by the cross section that is exposed to the Sun at any given time. As the planet rotates, this energy is distributed across the entire surface area of the globe.

For you to have a better feeling of the amount of energy that each part of Earth receives, think about taking a snapshot of Earth as it revolves. This snapshot would essentially freeze Earth's rotation for an instant. In that instant, only one-fourth of Earth is visible to the Sun. The amount of energy that this snapshot of Earth's cross sectional area would receive is one-fourth of the total given to Earth in one rotation. During the snapshot, it would have received the full 1366 watts per square meter. But some would be reflected. The clouds or terrestrial areas would absorb some, resulting in 51 percent, or 696 watts per square meter, being of use. Then, as Earth revolves during the other three-quarters of its rotation, it essentially receives no solar energy and is cooling and radiating heat energy away from Earth's surface. So, during the complete rotation, the energy averages out to 342 watts per square meter actually received if clouds do not block any solar energy.

At any given location and time, the amount of this energy that is actually received at Earth's surface depends on the state of the atmosphere (clouds). The latitude with the equator receives the highest amount. Compared to the other parts of Earth, this is why it is so hot at the equator. I will discuss how the heat of the Sun is distributed. The Sun emits abut two billion times (2×10^9) the amount of radiation that Earth catches. That is, about 3.86×10^{26} watts is radiated on a constant basis in all directions from the Sun's surface, most of it out into space.

Earth/Sun Energy Balance

For the sake of this literature, we can assume the Earth and Sun reached a status of energy neutral about a million years ago. Prior to reaching this point, the Sun was providing Earth with more energy than Earth was radiating back into space. This was especially true when Earth lost its first atmosphere and gained its second. This second atmosphere was mainly nitrogen, carbon dioxide, and water vapor, which resulted in the very dramatic greenhouse effect.

There was no oxygen present in the atmosphere at the time or very little. During this time, Earth maintained much of the energy received from the Sun. All of the Sun's radiation that Earth received was not reradiated out to space due to high cloud cover and the composition of the clouds and the atmosphere. The atmosphere was highly composed of water vapor and carbon dioxide. Much of the Sun's energy was absorbed in the oceans and other water concentrations. Earth radiated infrared wavelengths back toward outer space. The reradiation was mainly infrared and was captured by the water vapor, methane, and carbon dioxide in the atmosphere.[30] These gases then radiated in all directions, thus losing approximately fifty percent of this captured infrared radiation to space. But the portion reradiated down to Earth resulted in a fairly constant increase in Earth's temperature. The Earth warmed up considerably.

As Earth gained oxygen along with the growth of plants through photosynthesis, carbon dioxide was essentially eliminated. The Earth began to lose some of this stored heat energy. With the loss of carbon dioxide in the atmosphere, there was less infrared being absorbed. Therefore, there was less global warming. This continued for a significant amount of time. It may have taken until the Cambrian time frame, which was approximately 450 million years ago, for Earth to build up the oxygen and reduce the carbon dioxide to approximately today's levels. We may have reached this point around the time of the Cambian Explosion. At that time, it is estimated that the oxygen level was about 10 percent of the level it has reached over the last million years.

By today's levels, I am assuming we reached a point in time where Earth's radiation levels that escape to space equals the incoming Sun's radiated energy to Earth's surface. With this balance, Earth essentially retained the stable temperature we enjoy today. This is energy in

---

[30] There was no oxygen present in the atmosphere at the time or very little.

balance. To describe this status of energy neutral and give one a better understanding of my choice of words, see the following figure:[31]

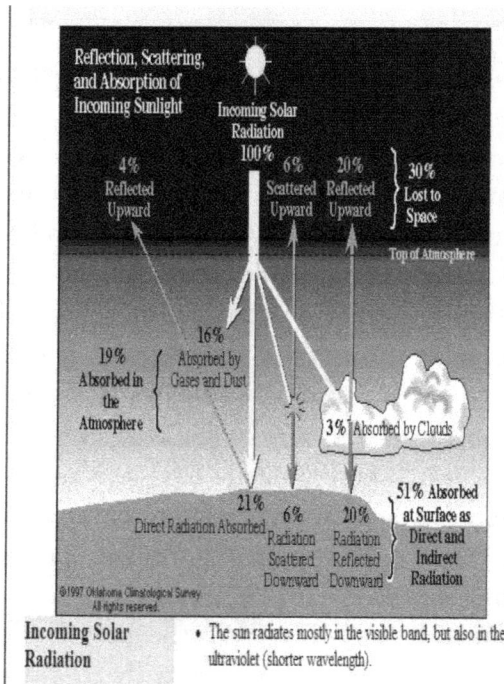

Figure 5. Incoming solar radiation.

The energy from the Sun is shown as being radiated with about 51 percent reaching the surface of Earth, 19 percent being absorbed in the atmosphere, and 30 percent lost to space via reflection from clouds.

Of the 51 percent that Earth's surface receives, 70 percent is reradiated back from the surface and clouds. It is lost to space or absorbed by the atmosphere, 23 percent is latent heat flux, and 7 percent is sensible heat flux that remains. The latent heat flux and sensible heat flux will be discussed later in the book. There is 21 percent of infrared from the surface that goes to the clouds or is absorbed by the water vapor or carbon dioxide in the atmosphere. This is not shown as staying on Earth because it is part of the infrared that is lost to space via

---

[31] physicalgeography.net/fundamentals/7j.html

clouds, water vapor, carbon dioxide, and methane after absorption by these medium. This is also the heat that can result in global warming.

This total amount of energy of the incoming and the outgoing energy is termed earth's energy budget. This is the result of the Sun's radiation energy to Earth and Earth's reradiation back toward space. This balance results in Earth's budget.

Earth's Energy Budget and the Atmosphere's Impact on Our Temperature

Absorption and reemission of radiation at Earth surface is only one part of the intricate web of heat transfer in Earth's planetary domain. Equally important are selective absorption and emission of radiation from molecules in the atmosphere. If Earth did not have an atmosphere, the surface temperature would be too cold to sustain life. If Earth had no atmosphere, the globally averaged surface temperature would be 0.4 degrees Fahrenheit (-18 degrees Celsius). Because Earth does have an atmosphere, the average surface temperature is actually 59 degrees Fahrenheit (15 degrees Celsius), so we have our atmosphere to thank for the comfort of this average temperature. I believe this should have been called global warmth rather than global warming, indicating it was at equilibrium and not warming further.[32] This has been constant for thousands of years.

The Earth's surface, oceans, humans, animals, atmosphere, and clouds emit radiation in the infrared band and near infrared band. Outgoing infrared radiation from Earth's surface, also called terrestrial radiation, is either lost to space or selectively absorbed by certain molecules, particularly water vapor, carbon dioxide, and methane. Gases that absorb infrared radiation are termed collectively as greenhouse gases.

---

[32] en.wikipedia.org/wiki/Evaporation; In *Semiconductor Devices: Physics and Technology* by Simon M. Sze, there is an especially detailed discussion of film deposition by evaporation; Martin A. Silberberg, *Chemistry*, 4th ed. (New York: McGraw-Hill, 2006).

Water vapor and carbon dioxide emit infrared radiation. Infrared radiation from greenhouse gases in the atmosphere is emitted in all directions, including back to Earth's surface. Approximately 25 to 50 percent can be radiated back to Earth. This reemission to Earth's surface maintains a higher temperature on our planet than what would be possible without the atmosphere. Interestingly, the clouds derive their heat from Earth's reradiation. The temperature of the lower part of the clouds provides temperature to Earth. As one would travel vertically up toward space, the temperature of the clouds falls rapidly. However, the various levels of our atmosphere change from cooling as the vapors rise to heating in some levels and then back to cooling. The clouds and precipitation produced from them provide the toughest challenge for the scientists who model our atmosphere in order to determine the climate that is in store for us. This is a tough challenge.

Water vapor in clouds is also an efficient absorber. It absorbs and radiates infrared radiation in all directions better than carbon dioxide, and there is more of it. These clouds act similarly to greenhouse gases to incoming radiation of the suns rays. Water vapor and carbon dioxide in the atmosphere also absorb some of the Sun's radiation. It is not a one-way street. So, if there is an increase in either one of these gases by any amount, it will increase the amount that the Sun's incoming infrared energy is absorbed as well as acting as greenhouse gases to earth's reradiation.

Satellite infrared imagery detects infrared emission from clouds and Earth's surface. It can be used during both day and night to determine the actual infrared being radiated away from Earth and into space. This is possible because sensitive electronic detection systems can detect infrared during day light and nighttime. The Sun's radiation exposes only one-fourth of Earth at any one instant during Earth's rotation. That fourth of Earth reradiates heat back during all twenty-four hours of its rotation. Therefore, the radiation on any spot on Earth is seen for about six hours. Then that part of Earth is in darkness or partial darkness and continues its reradiation during the whole twenty-four hours of its rotation.

We know how much energy is brought to Earth on a constant

basis as it spins its day away. It is much more difficult to determine how much is lost. It would be nice if we had a satellite located at four places in Earth's spin that gave us real-time measurements of Earth's gain and loss of heat. We could actually see how much is lost during any part of the day or night. As it is, we make some assumptions for this information and feel it resolves itself that we reemit all the energy that was received over the course of time back into space. There is a lag time in this scenario because the energy being emitted back into space does not represent the exact same energy that was received that day. It is much easier to calculate this over a longer time period. Because it is not an instant occurrence, this does not detract from the fact as long as it is continuous. Over time, many different ways as well as different people and organizations have measured this phenomenon. The agreement is within a few tenths of a percent.

When averaged over a year, the incoming energy to the atmosphere and Earth's surface equals the outgoing energy. This is the Sun/Earth energy balance. With this balance, Earth has essentially maintained the same climate over many years.

If we consider the entire earth-atmosphere system, then the amount of radiation entering the system must equal the amount leaving. If not, the system would continually heat or cool. This has not been the case. Thus, we are in true balance. Not all of this energy is radiative energy. Some is sensible and latent heat released indirectly from Earth.

Considering the atmosphere alone, the atmosphere experiences radiative cooling. This is the result of the atmosphere and clouds radiating heat energy back into space, resulting in its cooling. The atmosphere is kept from a net cooling by the addition of energy via latent and sensible heating.

Most of the Sun's radiation that passes through the atmosphere to hit Earth is in the visible part of the spectrum, that is, the part of the electromagnetic spectrum that is visible to the human eye. It is normally called simply light. This range of light we see has wavelengths that are 400 to 700 nanometers.

Most of the Earth's radiation that escapes the atmosphere or is absorbed by certain gases in the atmosphere are in the infrared band of wavelengths between 8,000 and 11,000 nanometers (This is 8 to 11 microns and a micron is 1/25000 of an inch). An example of the infrared is the heat that humans emit. Soldiers use special infrared goggles when they want to see the enemy at night. With these glasses, they can see the shape of human beings at night due to the infrared heat the body gives off. They can also see animals at night with these goggles for the same reason.

Wavelengths in the ultraviolet are shorter wavelengths than visible light. Ozone in the atmosphere absorbs a high percentage of them. Although mostly blocked by the ozone, these wavelengths cause one to get a sunburn or tan. These wavelengths are prevalent from 10:00 AM to 3:00 PM. People are warned to use special sunscreen lotion on their body to prevent the penetration of these wavelengths, especially during these hours.

Global Heat Balance

The equator receives much more heat than the North and South Poles. This difference in temperature results in the winds previously discussed. However, not all the heat absorbed by the area of the equator is transferred via the winds. Some is transferred by phenomena called heat fluxes.[33]

---

[33] en.wikipedia.org/wiki/Evaporation; In *Semiconductor Devices: Physics and Technology* by Simon M. Sze, there is an especially detailed discussion of film deposition by evaporation; Martin A. Silberberg, *Chemistry*, 4th ed. (New York: McGraw-Hill, 2006).

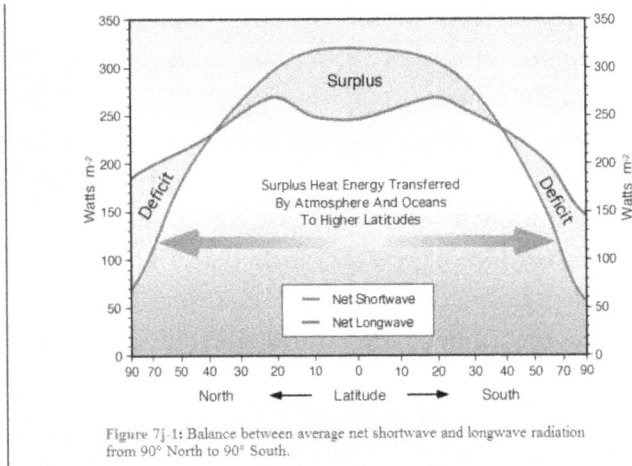

Figure 7j-1: Balance between average net shortwave and longwave radiation from 90° North to 90° South.

Figure 6. Balance between average net shortwave and longwave radiation from 90 degrees north to 90 degrees south.

This figure illustrates the annual values of net shortwave and net longwave radiation from the South Pole to the North Pole. On closer examination of this graph, note that the lines representing incoming and outgoing radiation do not have the same values for various parts of Earth. From 0 to 30 degrees latitude north and south, incoming solar radiation exceeds outgoing terrestrial radiation. A surplus of energy exists. The reverse holds true from 30 to 90 degrees latitude north and south. These regions have a deficit of energy. If this imbalance existed without transfer of the heat energy from around the equator and lack of heat around the poles, it would probably be impossible to live in certain areas of the world. However, Mother Nature has a way of partially balancing this problem. Surplus energy at low latitudes and a deficit at high latitudes result in energy transfer from the heat energy of the equator to the lower heat energy of the poles. This meridional transport of energy causes atmospheric and oceanic circulation as well as other fluxes to work toward a balance of temperature across the earth. Although it never balances it does a fair job of tempering the high and low temperatures across the globe.

A very high level of the sun's energy hits in the middle of our globe, representing the area around the equator, which receives about 350 watts per square meter. This level drops off dramatically as one goes away from the equator. However, the dip in the resultant energy drops the energy around the equator to approximately 250 watts per square meter. Observe how this energy is distributed on the curve to fill in the deficit in the parts of Earth in the temperate zone to around 200 watts per square meter. If there were no energy transfer, the poles would be 77 degrees Fahrenheit (25 degrees Celsius) cooler. The equator would be 57.2 degrees Fahrenheit (14 degrees Celsius) warmer. Obviously, these temperatures would have been uninhabitable at the equator and poles. This chart is another piece of evidence of how Mother Nature does her little tricks to keep things balanced on Earth.

Climate and the Weather

Weather is the great mixer for Earth. It handles the movement of the Sun's energy and the Earth's water, mixes the gases, and transfers much of the heat from one place to another on Earth or in the Earth's atmosphere. Weather is nature's means of trying to provide equilibrium. It is nature. The more you read, the more you will realize the part that weather plays in our harmonious lives here on Earth. I cover weather rather heavily in this discussion about global warming because of the major impact it would have on any climate change or global warming.

Several major phenomena determine the climate and weather on Earth. The first is the Earth's tilt. The Earth is tilted approximately 23.44 degrees with respect to the plane of the solar orbit. This tilt results in the seasons as Earth circles the Sun. This is the major reason for the climates we experience.[34] When the North Pole is tilted toward the Sun, the Northern Hemisphere will have summer. The Southern Hemisphere will have winter. This is the result of this half of Earth receiving warmer radiation as it is closer to the Sun due to our tilt and where we are in our circle of the Sun.

The opposite is true when the South Pole is pointed toward the Sun during Earth's obit. The Southern Hemisphere will have summer; the

---

[34] aoml.noaa.gov/hrd/tcfaq/D1.html

Northern Hemisphere will have winter. Midway in its orbit around the Sun, neither the North Pole nor the South Pole is pointed toward the Sun. There will be spring or autumn, depending on which side of the Sun the Earth is during its orbit.

A solstice is when Earth has its maximum tilt toward or away from the Sun. An equinox is when the tilt is minimized. Winter solstice occurs around December 21. Summer solstice is near June 21. Spring equinox is around March 20. Autumnal equinox is around September 23. However, this is for the Northern Hemisphere. The Southern Hemisphere is the opposite. Thus, the seasonal effects in the south are reversed. Fortunately, Earth has this tilt. If it was perpendicular to the plane of its orbit around the Sun, Earth would get too hot for human life.

Interestingly, Earth is spinning like a top. If you see a top spinning, it sort of wobbles in a circular pattern while maintaining its balance. The same is true of Earth. It has a very slow wobble that takes about 25,800 years to complete the circle of the wobble. This may have some effect on the climate.

In modern times, Earth's perihelion[35] occurs around January 3. The aphelion occurs around July 4. Coincidentally, the Southern Hemisphere is tilted toward the Sun at about the same time that Earth reaches the closest approach to the Sun, that is, perihelion. This results in about a 6.9 percent increase in solar energy reaching the Southern Hemisphere at perihelion. However, this effect is much less significant than the total energy change due to the axial tilt. The higher proportion of water in the Southern Hemisphere absorbs most of the excess energy. So, Mother Nature has helped again by increasing the size of the oceans on that part of Earth.

This discussion relates to the climates of Earth. Climates change slowly. In some cases, it is easier for scientists to project climates than weather. Climates work in trends, while weather can be disruptive from day to day.

Obviously, our planet takes a certain time to circle the Sun. Its tilt provides us the four seasons we enjoy as part of our climate. There

---

[35] Orbit takes it closest to the Sun

are other major causes for the weather on Earth, and they are due to Earth's shape. The Earth is not a perfect globe. It is much fatter at the equator, which is due partly to the gravitation pull between Earth and the moon. Even if there were no difference in the circumferences, the equator receives significantly more radiation than the North Pole or South Pole, which results in the equator being much warmer than the poles as discussed earlier. Now we will discuss how this affects the weather. Because energy in the form of heat tends to move from the hottest to the coldest areas, this is what happens with our planet. The heat from the equator tries to move toward the colder poles. This would result in wind moving from the equator toward the poles. However, at the same time, Earth is spinning at 1,000 miles an hour in a clockwise movement, which causes the winds to move from west to east. Because there is a vector of movement toward the poles and the spin of Earth provides a vector from west to east, the winds veer off at an angle from the equator. This is the major cause of winds that make up Earth's everyday weather. It is at least a large part of the everyday weather.

The dramatic contrast between the cold polar air and the tropical air from the equator gives rise to the jet stream. The in-between latitudes of the temperate zones have weather dictated by the instabilities of the jet stream airflow. The jet stream's path is also a function of the time of the year and the climate dictated by the position of Earth with respect to its orbit around the Sun. Because Earth is spinning, the jet stream follows a path that is like a sine wave oscillating in its path around the globe. This is caused by the puffs of hot air that are directed from the equator when it is exposed as it spins at 1,000 miles an hour. For about six hours, one portion of the equator is exposed and heats up. Then it tails off as another portion of Earth appears before the Sun. One can think of the Sun puffing at Earth, even though it is constant. I am probably the only one who calls this puffing, but it is partly to do with the jet stream circling the globe in an up-and-down motion. The move up begins as a portion of Earth is exposed. It reaches the top of this wave when it is directly exposed. Then it starts back down as it moves out of the direct path of the sunrays.

Different processes, like monsoons and organized thunderstorm systems, cause the weather systems in the tropics. On local levels, temperature differences can occur because different surfaces, like oceans, forests, ice sheets, and man-made objects, have different characteristics such as reflectivity, roughness, or moisture content. Surface temperature differences, in turn, cause pressure differences. A hot surface causes the air to rise and heats the air above it. The air expands, lowering the air pressure. The resulting horizontal pressure gradient accelerates the air from high to low pressure, creating local winds. The Earth's rotation then causes curvature of the flow via the Coriolis Effect. This path results in spiraling areas of high and low pressure. The high-pressure systems revolve in a clockwise rotation. The low-pressure systems revolve in counterclockwise rotation on the outside edges of the high pressure area. They are also smaller in size. This difference in size results like gears in a clock or car. The smaller low-pressure areas spin more rapidly as they move along the outer edge of the high-pressure system. Because of this, the low-pressure areas are very windy and usually bring stormy weather with them.

Because the Earth's axis is tilted relative to its orbital plane, sunlight is incident at different angles at different times of the year. In June, the Northern Hemisphere is tilted toward the Sun. At any given latitude in the Northern Hemisphere, sunlight falls more directly on that spot than in December, resulting in the warm climate of summer. Thus, this explains the seasons we experience.

Almost all familiar weather phenomena occur in the troposphere.[36] Weather does occur in the stratosphere and can affect weather lower down in the troposphere, but the exact mechanisms are still not completely understood or modeled. It is also difficult to obtain data of this huge part of our atmosphere, partly because the air is rare at these altitudes and the volume is so large. Meteorologists keep working and improving on computer programs to make predictions of the weather better and this includes determining how much the stratosphere affects our climate and our weather.

Weather is one of the fundamental processes that shape the Earth.

---

[36] The lower part of the atmosphere

The weathering process breaks down rocks and soils into smaller fragments. Then they are broke down even further into their constituent substances. This results in variations of the oxidation and other chemical reactions of these smaller increments due to the increase in their surface area. Weather is a form of energy, whether it is in the form of a change in temperature, surface erosion, winds, and changes in water content of Earth's surface or destruction of other structures on the face of Earth.

Sensible Heat Flux, Latent Heat Flux, and Surface Heat Flux

The redistribution of energy across the Earth's surface is accomplished primarily through three processes: sensible heat flux (direct), latent heat flux (evaporation), and surface heat flux (flow of currents in the ocean).[37]

Sensible heat (direct) flux is the process where heat energy is transferred from the Earth's surface to the atmosphere directly by conduction and convection. This energy is then moved from the tropics to the poles by advection, creating atmospheric circulation. As a result, atmospheric circulation moves warm tropical air to the Polar Regions and cold air from the poles toward the equator. Nations in the temperate zones, for example, the United States, benefit by this air movement from the equator and the poles by having a climate that is temperate and is not subject to extreme hot or cold weather. Much of the heat energy is in the infrared wavelengths. The oceans do absorb most of it; mainly at the surface of the water. As a result, Earth's oceans are warmer than the air above it by a little less than about 2 degrees Fahrenheit (1 degree Celsius). Direct heat transfer occurs from the water to the air by conduction, via infrared wavelengths, because it takes much less energy to heat air than to heat water. The heat energy required to increase the temperature of water that is 1.0 centimeter (0.4 inches) deep is enough heat energy to raise the temperature of 31 meters (approximately 100 feet) of air above it by about 2 degrees Fahrenheit (1 degree Celsius).

---

[37] climatesci.colorado.edu/2007/04/05/evaporation-is-equal-to-precipitation-on… 7/26/2007

This heat energy is transferred in the infrared wavelengths. It causes turbulence in the air, which results in this warm air rising and exposes the cool remaining surface water; thus allowing additional heating by the sun of the cool surface water of the oceans to start the cycle over again.

Heat input into the atmosphere from below cause's instability via a reduction of density, which results in atmospheric convection and turbulent and the upward transport of heat. This kind of mechanism starts hurricanes. In contrast, heat input into the ocean from above increases stability through reduction of density at the surface. It prevents efficient heat penetration into the deep layers of the ocean.

Latent heat flux moves energy globally when solid (ice) and liquid water of the oceans are converted into vapor (evaporation). This vapor is often moved by atmospheric circulation vertically and horizontally to cooler locations where it is cooled by the cool air of the clouds and is condensed as rain or deposited as snow, releasing the heat energy stored within it to the atmosphere above. Thus the heat lost by the oceans surface or ice via evaporation is transferred to the atmosphere during the condensation and resulting precipitation. This heat that is transferred to the atmosphere escapes to space. This completes this cycle of the suns heat that strikes the oceans or ice being released back to space and conserves the balance of energy.

Latent heat flux is associated with evaporation of water at the surface, mainly by the oceans, and subsequent condensation of water vapor in the troposphere. It is an important component of Earth's surface energy budget. This is a form of air-conditioning that Mother Nature provides. The heat from the Sun causes huge amounts of evaporation from the surface of the oceans, lakes, and rivers. This evaporation of water is taken into the clouds. The same process that creates winds and weather changes moves it. The condensation results in the water that was evaporated by the heat of the Sun in the oceans to be carried to various locations in the world and dropped in the form of precipitation. As a result, two major phenomena occur: the transfer of heat to the atmosphere as the water vapor is condensed and the transfer of potable water to all parts of Earth. Through this mechanism, drinkable water is transferred in large quantities to various parts of the world. The evaporation of water from the oceans leaves the salt behind. It is Earth's desalination system. Our Earth can provide freshwater to oxygen-breathing animals, human beings, and to the growing plants.

Finally, large quantities of radiation energy are transferred into the Earth's tropical oceans. The energy enters these water bodies at the surface, where absorbed radiation heat energy is converted into the water's heat energy. Conduction and convection transfer the warmed surface water, which does not evaporate, downward into the water column. Heated water goes the same route as heated air. It moves from the warm waters to the colder waters. These water currents are another means that Mother Nature has provided to spread the sun's heat energy to the colder waters and thus warm the land that is adjacent to this flow of warm water. The Gulf Stream is a typical example of this means of ocean water transferring heat to the colder waters and to warm countries; and example of this is the warming of England. Solar energy received by the ocean varies irregularly with wavelength.

The percent absorption of the heat energy in the sea reduces dramatically with water depth: [38]

- 73 percent reach 1.0 centimeter (0.4 inches)
- 45 percent penetrate to 1.0 meter (39.4 inches)
- 22 percent penetrate to 10 meters (32.8 feet)
- 0.53 percent penetrate to 100 meters (328 feet)
- 0.0062 percent penetrate to 200 meters (656 feet)

The minimum energy supply necessary to maintain photosynthesis is 0.003 cal cm^{-2}. Thus, photosynthesis is supported at 200 meters. Under optimum conditions, that is, absolutely clear water, this amount is available at 220 meters (722 feet).[39] The ocean's plant life provides almost half of Earth's oxygen via photosynthesis.

Horizontal transfer of this heat energy from the equator to the poles

[38] climatesci.colorado.edu/2007/04/05/evaporation-is-equal-to-precipitation-on... 7/26/2007

[39] climatesci.colorado.edu/2007/04/05/evaporation-is-equal-to-precipitation-on... 7/26/2007

is accomplished by ocean currents. One example is the Gulf Stream, where water is heated in the Gulf of Mexico and streams past Florida and along the eastern part of the United States. As a result, the Gulf Streams warms the waters of the Atlantic Ocean. This warming extends all the way up to the waters around the North Pole. The Scandinavian countries benefit from this warming. It helps them to remain fairly comfortable, considering the latitude where these countries are located.

The ocean/atmosphere/Sun heat balance is similar to the one described between Earth and the Sun. The heat energy that the Sun provides to Earth's water systems is equal to the losses mentioned previously. It is expressed in a balanced equation by net radiative heat gain minus evaporative heat loss minus direct heat loss equals zero.

This condition has existed for many years. If not, the ocean's temperature at the surface and below would have changed. The energy of the radiation is proportional to the fourth power of the absolute temperature.[40] In this equation, the Sun's temperature is 5780 degrees Kelvin and Earth is 279 degrees Kelvin. Thus, daily or seasonal variation in the ocean's surface temperature of a few degrees Celsius, the same as Kelvin, has little effect on the back radiation energy because these variations are small compared to the absolute Kelvin temperature level. Isn't Mother Nature grand?

Evaporation

It is important to review evaporation in order to better understand this means of heat transfer and overcome some misconceptions. Vaporization and evaporation are sometimes misunderstood and thought to be the same phenomenon. Evaporation is the process whereby atoms or molecules in a liquid state gain sufficient energy to enter the gaseous state.[41] It is the opposite process of condensation. Evaporation is exclusively a surface phenomenon, and it should not be confused with boiling, or vaporization. Most notably, for a liquid to boil, its vapor

---

[40] Stefan-Boltzman Law, when expressed in temperatures of Kelvin
[41] In solids, the equivalent process is known as sublimation.

pressure must equal the ambient pressure. For water this is 212 F (100C). For evaporation to occur, this is not the case.

Evaporation is a critical component of the water cycle, which is responsible for clouds and rain. Solar energy drives evaporation of water from oceans, lakes, moisture in the soil, and other sources of water. In hydrology, evaporation and transpiration, which involves evaporation within plant stomata, are collectively termed evapotranspiration.

For molecules of the ocean to evaporate, they must be located near the surface and move in the proper direction, that is, not downward. They must have sufficient kinetic energy to overcome liquid-phase intermolecular forces.[42] Only a small proportion of the molecules meet these criteria, so the rate of evaporation is limited. Because the kinetic energy of a molecule is related to its temperature,[43] evaporation proceeds more quickly at higher temperatures. As the faster-moving molecules escape, the remaining molecules have lower average kinetic energy. The liquid's temperature decreases.

If the Sun did not evaporate the surface water from the oceans, they would gain in temperature from day to day and be uninhabitable. The act of evaporation is always a cooling phenomenon. It performs this cooling on the world's water systems. This phenomenon is also called evaporative cooling. This is why evaporating sweat cools the human body. Evaporative cooling is a cheap means of air condition and is used in areas like Arizona. Evaporative air conditioning is only effective if the environment has a low humidity; that's why it is effective in places like Arizona.

Evaporation also tends to proceed more quickly with higher flow rates between the gaseous and liquid phase and in liquids with higher vapor pressure. Knowing the highest energy particles are located at the surface of any liquid, evaporation occurs only at the surface of any liquid. One might believe that evaporation and boiling have the same chemical characteristics, but this is not true.

Evaporation is a surface phenomenon that occurs at all temperatures.

---

[42] ajcn.org/cgi/content/full/78/3/660S

[43] In fact, the kinetic energy determines the related temperature.

The amount of kinetic energy, or heat, applied to a liquid's surface determines how many particles can evaporate. Because evaporation is a variable process, it is hard to model and predict its complete impact on Earth's temperatures at any given time. The combination of a variable evaporation rate, the unpredictable consequences of weather and how it reacts to this variable rate, and the variability of heat transfer via precipitation makes weather difficult to predict. It may also make prediction of climate changes difficult. I have reviewed evaporation very thoroughly in order for you to see the type of forces that must be overcome in order for global warming to occur.

On the other hand, vaporization or boiling occurs throughout a liquid. It has a definite temperature, or boiling point, for each and every liquid. For example, it is 100 degrees Celsius (212 F) for water. Another significant difference from evaporation is that the temperature of a liquid stays constant throughout during vaporization.

Even at cool temperatures, a liquid can still evaporate, but only a few particles would separate from the liquid over a long period. So, the temperature is always changing during evaporation. There is another consideration for evaporation as it relates to climate, weather, atmosphere, and air. This consideration depends on the air itself. Air is like a big sponge that wants to take in as much water as it can hold at a given temperature. If it is not saturated with water vapor for its given temperature, it tries to take in more water vapor. It does not care if the temperature of the water is very high. If the kinetic energy of the water's surface is high enough, the air will swallow up the top surface of the water. If heat energy continues to be supplied to the water, the air will take in additional surface water vapor while cooling the surface it left. This is the removal of heat from the water's surface to clouds above in the form of water vapor. This water vapor is transported to other locations by the atmosphere. As it is cooled and condensed, it results in some form of precipitation. The clouds take on the heat of condensation. Isn't Mother Nature wonderful?

The Ocean Mass Water Budget

Obviously, the Sun's energy spills a lot of its energy on the oceans and waterways of the world. I have discussed the various moves of the energy from the ocean in different forms. One very important energy disposition occurs between the waterways and the solid ground of the world that is required for life forms. This is fresh drinking water. The removal of this freshwater and its return is called ocean mass budget, which results in a very vivid picture of Earth's circulation of water.

A large amount of water is relieved from the oceans through various means. It is in a different form than the water it vacated. This is a direct relationship with the Sun's energy heating the water. The ocean, of course, is salty, but the water that is removed via the methods discussed is fresh water. It is the only true source of the world's drinkable water. Just set back and close your eyes for a minute and picture what I am about to portray.

Nature's Big Circle (Hydrologic Cycle)

Think of a big circle that goes from the oceans to the clouds, to the land, and back to the ocean. This circle is the path of the salt-relieved water that is evaporated into the air and makes its way to various parts of Earth. It is condensed, resulting in precipitation like rainfall or snow showers. Now picture this freshwater from the precipitation flowing in the rivers and lakes of the world. Much of it is used by humans, animals, and plants to provide the necessary water requirement for life. It is water flowing from the mountains when the snow melts. It fills the wells in Earth that flow up to the creeks and to the rivers. Its water fills the aquifers underground that flow toward other water outlets, being filtered by Earth's filtering system as it moves on its way through the aquifers. It is water flowing from the melting snow or rain and filling the reservoirs. It is rivers that fill and flow from the precipitation or the melting of the snow. It is freshwater lakes that fill from the precipitation while being supplemented by the flow of the rivers into and out of it.

This journey of the water via the waterways of Earth provides the world it's only source of this freshwater. Eventually, the flow of the rivers finds their way back into the oceans, thus completing this fantastic journey and completing this circle that is required for life. Isn't Mother Nature wonderful?

Just think of this circle and how unique it is. The ocean's waters are desalinated through evaporation to provide this huge circle. The rivers of the world bring back the unused water to the oceans. If things were perfect, the oceans would look like they did before this cycle began, that is, they would have their same salinity. That's what completes this natural cycle.

This is the evaporation-precipitation cycle, and it is quantified.[44] It takes 585 Calories of the Sun's energy to convert a gram of water into a vapor. It takes only 80 Calories of the Sun's energy to convert ice to water at 32 degrees Fahrenheit (0 degrees Celsius). Likewise, converting the water vapor back into water requires the removal of 585 Calories per gram during condensation.[45] (This is why steam engines worked so well. Steam was generated by burning wood or coal and then received these Calories back as the steam did its work and lost the 585 Calories for each gram returned to the water state.) It only takes the removal of 80 Calories per gram to convert water at 32 degrees Fahrenheit (0 degrees Celsius) to ice.

The equation for the ocean mass budget is evaporation minus precipitation equals zero. The evaporation must equal the precipitation. In order to determine if this equation is holding true, the salinity of the ocean's surface waters must be tested. The salinity of the ocean's surface water should remain constant over time, and this has been the case for many years. The precipitation has to be taken from observations over the world. The highest precipitation is just north of the equator and at 50 degrees, plus and minus, where the polar fronts exist.

This cycle has a delay built into it. The clouds of Earth can only hold water for eight or nine days before there is some form of precipitation.

---

[44] ajcn.org/cgi/content/full/78/3/660S
[45] in the clouds that take up the energy of condensation

It takes time for the water that is evaporated to make this cycle of evaporation and precipitation and flow of water in the various waterways. Then it takes time for the precipitation to flow down the rivers and other canals to complete its journey back into the ocean. Therefore, the precipitation is not directly related to the evaporation in this equation for a given time or day. However, this is not a problem as long as the cycle is a continuous one. In this case, the return of water to the ocean is from evaporation from a previous time. This equation of ocean mass budget represents a snapshot that should be continuous and checked at the same time each year. As long as this holds true, one can consider the circle happening in the same day and on a continuous basis.

The ocean mass budget involves the effects of evaporation and precipitation on the amount of water in the ocean. The effect on the total amount of water is significant only on a geological time scale. This is true based on the fact that there is significant delay from the time of evaporation to the time of rivers feeding freshwater back into the ocean. The major issue is whether the cycle results in the change of the salinity of the ocean's water. For the evaporation and precipitation of freshwater along with its flow back into the ocean to match, the salinity of the ocean before the evaporation must equal the salinity of the ocean after the flow of freshwater back into the ocean. In other words, evaporation takes out fresh water, and the flow of freshwater back into the ocean must be equal for the salinity to match. The fact that measurements have shown no significant change in the salinity of the ocean's surface water is proof that the water cycle is continuous and real time ... and perfectly balanced. If there were the possibility of further warming of Earth as a result of additional global warming, this cycle must be maintained. If it is maintained, it will result in an increase in evaporation; the freshwater to be provided to the animals, plants and humans on Earth; and the flow of water back into the oceans.

This is a subject of much discussion since many factors, including the oceans, evaporation, the amount of water the clouds can hold, the number of days of precipitation the clouds can hold, and this additional amount that the cycle is trying to accomplish to keep Earth at a constant climate, are all variables that result in demands on the system that is presently in balance. Scientists are debating this subject. Additional attempts of climate change must result in some change in accommodating this circle. It is possible that larger clouds could form and act as reflectors to the Sun's energy and prevent additional heating of Earth, thus offsetting any change in climate. There is also the possibility that the additional water vapor and carbon dioxide in the atmosphere would result in the absorption of additional infrared heat coming in from the Sun and return half of it back toward space, thus offsetting any attempt at global warming.

The oceans of the world provide the world with a buffer for any increase in climate. It is the key to prevent any additional changes in the climate. This is a huge enertia that would need to be overcome Mother Nature's efficient cycle is probably the most difficult function on Earth to change.

The distribution of ocean's surface salinity mirrors the distribution of the evaporation-precipitation cycle over large parts of the ocean. Deviations occur from river runoff. On a global scale, the balance is: (42)

Evaporation	=	440×10^3 km^3 year^{-1}
Precipitation	=	411×10^3 km^3 year^{-1}
River runoff	=	29×10^3 km^3 year^{-1}

Except on the geological time scale, the melting and freezing of ice is balanced. Most rivers are found in the Northern Hemisphere, so the proportionality between sea surface salinity and evaporation-precipitation is better over most of the Southern Hemisphere.

The average time for evaporated water to reside in the atmosphere in a cloud formation is eight to nine days. So, the water that is evaporated today will result in some form of precipitation in a little over a week. During this time, the winds and other elements can move these clouds and distribute this water, snow, or other form of precipitation to Earth. If you consider this time before there is precipitation, then it turns out that the clouds empty their precipitation an average of forty-five times a year.[46] This happens at various points across the globe. Therefore, we find times when there is a large amount of precipitation in some area of the world while other areas are suffering from drought. So, the atmosphere acts like several big jugs of water for the evaporation collection before precipitation occurs. Any increase in the evaporation rate due to global warming will result in a decrease in the number of days that the water is held in the atmosphere and an increase in the annual precipitation. The atmosphere can only hold so much water. It is like a sponge that can fill so much. Then it gets squeezed out.

This is only true if all things remain constant in these calculations. If there is a change, for example, if the cloud formations are bigger, the result could be additional reflection of the Sun's energy and keep the surface water from heating. By reviewing the amount of rainfall in the world annually, we can determine if there is any increase. If global warming were occurring, we would expect to see an increase in the yearly rainfall. To date, this has not been the case.

The amount of precipitation may change its location, but the total amount does not change over a period of time. Each year, there are areas in the world with abnormal rainfall, either increasing or decreasing. This is obvious based on the many media stories that are available. For instance, there is a flood situation in Texas as I write this book. Last year, there were floods in other parts of the United States and other parts of the world. At the same time, there are media reports of the lack of rainfall in certain parts of the world each year. This year, for example, it happens to be in Australia. Other places in the world are suffering from drought at the same time that other places are being flooded. In some cases, the lack of rain in a location over a period of time results in a drought, and it is treated as a serious situation. I live just about ten miles from San Jose, California. This morning's newspaper reported that San Jose only received 9.28 inches of rainfall in the 2006–2007 rain years. This represents approximately 62 percent of normal. However, San Jose's driest year was in 1877 when 4.83 inches of rain fell. In 1989, San Jose had a rainfall of 8.32 inches. The most rain fell in 1890 when

[46] Arizona Cooperative Extension, College of Agriculture, Tucson, Arizona

30.30 inches of rain was recorded. Likewise, 30.25 inches fell in 1983. (San Jose weather information from the San Jose Mercury News)

This gives one the idea of how random some rainfall events are. This recent dry spell comes after more than a decade of healthy rainfall. Clearly, the media exposes the lack or abundance of rain. Then the media relates it more recently to global warming, which may or may not be true. That's the purpose of this book; to see if what is being seen is a climate change, a change due to man's pollution of our atmosphere with carbon dioxide, or some other phenomenon. But the media's expounding on this subject has brought it to the attention of more people. Many people are now scared of the effects that global warming may have on them personally.

Over the past several thousand years, the Sun has placed a constant amount of energy on Earth, year by year. At the same time, Earth's ocean levels have remained fairly constant. So we have this constant energy, this constant large amount of water, and a yearly rainfall that has remained constant all over the world. Changing this requires a huge action of nature to be able to overcome the large inertia that the world's large bodies of water provide. Global warming must overcome this to be a fact.

Chemical Composition (Background Data)

Man is now about to enter into the picture. At this time, we will summarize the status of Earth as man found it. This will give us a background from which to work and review any changes with man's presence.

The Earth's mass is approximately 6.0×10^{24} kilograms (about 1.32×10^{25} pounds). It is composed mostly of iron (32.1 percent), oxygen (30.1 percent), silicon (15.1 percent), magnesium (13.9 percent), sulfur (2.9 percent), nickel (1.8 percent), calcium (1.5 percent), and aluminum (1.4 percent). The remaining 1.2 percent consists of trace amounts of other elements.[47] Water covers 70.8 percent of Earth' surface. Of this number, 97 percent is salt water in the oceans and seas, 2 percent is freshwater lakes and rivers, 1 percent is snow and ice as glaciers, and 0.0570 percent (570 to 1000 parts per million) is atmospheric water. (43)

Oxygen

When reviewing Earth's past and present to determine what may be causing global warming, one has to understand the elements in the atmosphere and how carbon dioxide or water vapor may react with them. We have just completed the first phase of my hunt for what may be causing any global warming or climate shift. We saw an atmosphere forming that is directed toward what we have today. We will continue to review the atmosphere as it seems to have the clue to what may be global warming in its infancy. No matter what form global warming takes, it must include Earth's atmosphere and the large bodies of water.

When one considers the sun and how it provides all the energy for Earth, we also have to consider how Earth married this warm energy source and provided the necessary ingredients to produce oxygen. Without oxygen, there is no life as we know it. Oxygen is the most common component of the Earth's crust (49 percent by mass). It is the second most common component of the Earth as a whole (28 percent by mass). It is the most common component of the world's oceans (86 percent by mass), and the second most common component of the Earth's atmosphere (20.947 percent by volume), second to nitrogen.[48] The more common rock constituents of Earth's crust are nearly all oxides with chlorine, sulfur, and fluorine being the only important exception to

---

[47] en.wikipedia.org/wiki/Oxygen_cycle; Steve Nadis, "The Cells That Rule the Seas," *Scientific American* (Nov. 2003); P. Cloud and A. Gibor, "The oxygen cycle," *Scientific American* (September 1970): 110–123; J. Fasullo, "Substitute Lectures for ATOC 3600: Principles of Climate, Lectures on the global oxygen cycle," http://paos.colorado.edu/~fasullo/pjw_class/oxygencycle.html

[48] en.wikipedia.org/wiki/origin of water on Earth; Jörn Müller and Harald Lesch, "Woher kommt das Wasser der Erde? - Urgaswolke oder Meteoriten, " *Chemie in unserer Zeit* 37, no. 4 (2003): 242–246.

this. Their total amount in any rock is usually much less than 1 percent. The principal oxides are silica, alumina, lime, magnesia, sodium, iron oxides, water, titanium, and phosphorus pentoxide.(It is worth injecting a comment here about Oxygen that is beyond comprehension when you realize how technically competent man is relative to understanding many of the chemical reactions in the world of chemistry and biology today. It was not known until just before the Civil War (1860 AD) in the United States that the gas we breathe and is required by man to live is Oxygen).

In nature, the light-driven splitting of water during oxygenic photosynthesis in cyanobacteria, green algae, and plants produces free oxygen. Algae produce about 73 to 87 percent of the net global production of oxygen, which makes it available to humans and other animals for respiration. Another major source of oxygen is trees. Trees can absorb carbon dioxide at the rate of 26 pounds per year, especially young trees that are still growing, while releasing oxygen into the air. Young trees produce more oxygen and reduce more carbon dioxide than older trees. Therefore, the replanting of young trees to replace older trees cut down for its wood does not have to be done on a "one for one" basis. There are valuable programs that strive for the replanting of trees.

The oxygen cycle is the biogeochemical cycle that describes the movement of oxygen within and between its three main reservoirs: the atmosphere, biosphere (includes where we live and all animals live. This would include the atmosphere, and the water where other forms of life exist), and lithosphere (Earth's surface).

The main driving factor of the oxygen cycle is photosynthesis, which is responsible for our modern atmosphere and life as we know it. Because of the vast amounts of oxygen in the atmosphere, even if all photosynthesis was to cease, which is unlikely, it would take 5,000 to 2.5 million years to strip out more or less all of the oxygen. (44)

By far, the largest reservoir of earth's oxygen is within the silicate and oxide minerals of the crust and mantle (99.5 percent). Only a small fraction has been released as free oxygen to the biosphere (0.01 percent) and atmosphere (0.49 percent). (45)

The main source of oxygen within the biosphere and atmosphere is photosynthesis, which breaks down carbon dioxide and water to create sugars and oxygen:[49]

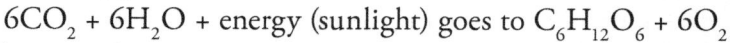

$$6CO_2 + 6H_2O + \text{energy (sunlight) goes to } C_6H_{12}O_6 + 6O_2$$

Photosynthesizing organisms include the plant life of the land areas as well as the phytoplankton of the oceans. The tiny marine cyanobacteria prochlorococcus was discovered in 1986 and accounts for more than half of the photosynthesis of the open ocean.[50] They are tiny, but a tremendous amount of them are in the oceans.

An additional source of atmospheric oxygen comes from photolysis, whereby high-energy ultraviolet radiation breaks down atmospheric water and nitrite into component atoms. The free hydrogen and nitrogen atoms escape into space, leaving oxygen in the atmosphere.

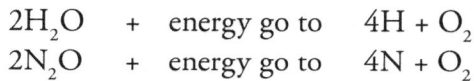

$$2H_2O \quad + \quad \text{energy go to} \quad 4H + O_2$$
$$2N_2O \quad + \quad \text{energy go to} \quad 4N + O_2$$

Mainly, oxygen is exchanged from the atmosphere via respiration of animals and humans and decay mechanisms in which animal, human life, and bacteria consume oxygen and release carbon dioxide. During this phase of the study, we will try to determine the effects of the major increase in man and animals. As these two increases in number, there is a greater generation of carbon dioxide and a greater demand for oxygen. As mentioned earlier, with 6.5 billion humans on Earth and many more animals than humans, there is a significant demand for oxygen and a search for ways to discard carbon dioxide. Decay of tree branches and leafs in forests that are overcrowded, thus not allowing sunlight to hit these materials that fall to Earth causes them to rot, is a reduction process that generates carbon dioxide. Much of this is occurring in several bands of forests in the world. Fossils show that many of these actions occurred early in the life of Earth and gave evidence that Earth was basically a reduction system versus today's oxidation system.

---

[49] en.wikipedia.org/wiki/Weather; Cynthia M. O'Carroll, "Weather Forecasters May Look Sky-high for Answers," Goddard Space Flight Center (NASA), October 28, 2001.
[50] Ibid.

Oxygen is also cycled between the biosphere and lithosphere. Marine organisms in the biosphere create calcium carbonate shell material ($CaCO_3$) that is rich in oxygen. When the organism dies, its shell is deposited on the shallow sea floor. Over time, as it is buried, the limestone rock of the lithosphere is created. Weathering processes initiated by organisms can also free oxygen from the lithosphere. Plants and animals extract nutrient minerals from rocks and release oxygen in the process.[51]

Oxygen Reservoir Capacities and Fluxes

The following tables offer estimates of oxygen cycle reservoir capacities and fluxes. These numbers are based primarily on estimates from J.C.G. Walker: *

Reservoir	Capacity (kgO$_2$)	Flux In/Out (KgO2 per year)	Residence Time (Years)
Atmosphere	1.4×10^{18}	$30,000 \times 10^{10}$	4,500
Biosphere	1.6×10^{16}	$30,000 \times 10^{10}$	50
Lithosphere	2.9×10^{20}	60×10^{10}	500,000,000

Table 3. Major reservoirs involved in the oxygen cycle from J.C.G. Walker.

[51] Scientists did not always know that the gas we breathe is oxygen. It was not recognized until just before the Civil War in the United States. Chart From J.C.G. Walker

Gains	
Photosynthesis (land)	16,500
Photosynthesis (ocean)	13,500
Photosynthesis of N_2O	1.3
Photosynthesis of H_2O	0.03
Total Gains	**Approximately 30,000**
Losses (Respiration and Decay)	
Aerobic Respiration	23,000
Microbial Oxidation	5,100
Combustion of Fossil Fuel (anthropologic)	1,200
Photochemical Oxidation	600
Fixation of N_2 by Lightning	12
Fixation of N_2 by Industry (anthropologic)	10
Oxidation of Volcanic Gases	5
Total Losses	**Approximately 29,927**
Losses (Weathering)	
Chemical Weathering	50
Surface Reaction of O_3	12
Total Losses	**62**
Total Losses (Respiration, Decay, and Weathering)	**Approximately 29,989**

Table 4. Annual gain and loss of atmospheric oxygen (Units of 1010 kilograms of O_2 per year). (46)

This is fairly old data, but it probably has not changed much.[52]

Water the World Over

Earth is covered by water to a great extent. 70.8% of Earth is covered by water; with 97% in oceans and seas, 2% in lakes and rivers, 1%

---

[52] Geolor.com/geoteach/How_Did_Earths_Atmosphere_Evolve-geoteach

snow and ice as glaciers and 0.057% is atmospheric water (570 parts per million).Water is not distributed evenly over Earth. In the Northern Hemisphere, the ratio of water to land is one-and-a-half to one. In the Southern Hemisphere, the ratio is four to one. The Pacific Ocean is the largest and deepest. The Atlantic Ocean is the second largest, but the Indian Ocean is deeper.[53]

About 97.5 percent of the water is saline. The remaining 2.5 percent is fresh water. The majority of the freshwater, about 68.7 percent, is currently in the form of ice. (47)

About 3.5 percent of the total mass of the oceans consists of salt. Most of this salt was released from volcanic activity or extracted from cool, igneous rocks. (48) If we could find a way to convert salt water into freshwater, we would solve the short supply of freshwater and probably resolve global warming. The oceans are also a reservoir of dissolved atmospheric gases, which are essential for the survival of many aquatic life forms.

The oceans are a heat reservoir for the planet. They are so large that drastic changes in climate must get over this huge entity first along with the inertia and time factor associated with any change. The change would have to take a tremendous amount of time, based on the consistency with which Earth has been supplied its energy and its daily release back of the energy. This is an important point related to our subject matter concerning any climate changes or global warming. The Earth retains its present climate because the energy that the Sun provides is a constant. This energy is returned to outer space each day to offset any changes in climate. In fact, this transfer of heat back from Earth keeps the temperature of Earth at a relatively constant temperature. On the other hand, the oceans are the most prominent weather factor on Earth and can cause weather changes rapidly. But we must remember that changes in weather are not changes in climate.

Sometimes, I think we should be thankful for hurricanes. They politely pick up only the freshwater from the oceans and leave the salt behind. Then they deposit the freshwater on land. This is true for

---

[53] math.ucr.edu/home/baez/temperature/

rain, snow, and any other form of precipitation. Weather is our only true provider of freshwater. Perhaps some bright engineer will invent a machine that simulates the hurricane effect and pulls the freshwater out of the ocean and leave the salt behind. Desalination using present methods is too expensive.

Energy of a Hurricane

While reviewing weather and the large actions of hurricanes earlier, I felt it would be worth spending additional time on this subject to provide information that might not have anything to do with global warming. However, if there is global warming, it will surely involve more hurricanes. Any additional heating of the world's oceans must result in increasing the number and perhaps the energy of each hurricane. Hurricanes represent the major release of any energy buildup of Earth's temperature. Not only do they release the heat of the ocean's surface, they also provide relief of the heating of the landmasses through precipitation. Hurricanes represent the biggest movers of freshwater to all parts of Earth. They provide significant cooling as well. Hurricanes are a major force on Earth. A hurricane brings destruction, but it also brings relief from drought. [54]

Initially, one can think of hurricanes as a heat engine. They obtain their heat input from the warm, humid air over the tropical oceans. Then they release this heat through the condensation of water vapor into water droplets in deep thunderstorm clouds. Then they emit a cold exhaust in the upper level of the troposphere, which is approximately 8 miles (12 kilometers) up. One can analyze the energies of a hurricane in two ways:[55]

---

[54] In some parts of the world, hurricanes are called typhoons.
[55] interactive2.usgs.gov/faq/listfaqbycategory/get answer.aspp?id=187

Method 1	The total amount of energy released by the condensation of water droplets	The vast majority of the heat released in the condensation process causes rising motions in the thunderstorms. Only a small portion drives the storm's horizontal winds.
Method 2	The amount of kinetic energy generated to maintain the hurricane's strong, swirling winds[56]	

Method 1	Total energy released through cloud/rain formation by water condensation	An average hurricane produces 0.6 inches (1.5 centimeters) per day of rain inside a circle of radius of 415 miles (665 kilometers).[57] This means that more rain falls in the inner portion of a hurricane around the eye and less in the outer rainbands. This converts to a volume of rain at 2.1×10^{16} cm^3 per day. A cubic centimeter of rain weighs 1.0 gram. Using the latent heat of condensation, this amount of rain produced gives 5.2×10^{19} joules per day or 6.0×10^{14} watts (1.43×10^{11} Calories per day). This is the equivalent to 200 times the worldwide electrical generating capacity, an incredible amount of energy produced!

---

[56] Emanuel 1999.

[57] Gray 1981.

Method 2	Total kinetic energy (wind energy) generated	For a mature hurricane, the amount of kinetic energy generated is equal to that being dissipated due to friction. The dissipation rate per unit area is air density times the drag coefficient times the wind speed cubed.[58] One could either integrate a typical wind profile over a range of radii from the hurricane's center to the outer radius encompassing the storm or assume an average wind speed for the inner core of the hurricane. Doing the latter and using 90 mile per hour (40 meters per second) winds on a scale of radius of 40 nautical miles (60 kilometers), one gets a wind dissipation rate (wind generation rate) of 1.3 x 10^{17} joules per day or 1.5 x 10^{12} watts (3.6 x 10^8 Calories). This is equivalent to about half the worldwide electrical generating capacity, also an amazing amount of energy being produced

From the two examples above, one can see from the first method that the amount of energy released in a hurricane by creating clouds and rain is 400 times greater than maintaining the winds, so most of

[58] For details, see Emanuel 1999.

the energy is dispensed in an area that humans do not witness. They just see the wind and rain, which is 0.0025 percent of the main energy.

However, this vertical movement up the funnel of the hurricane causes weather to occur in other places on Earth. Any movement of water vapor due to evaporation up to the clouds results in the eventual release of energy to the clouds. Condensation occurs wherever these clouds have moved to. These hurricanes can cause major rainstorms that occur thousands of miles away from the hurricane.

Comparing the Hurricane Energy to an Atomic Bomb

I thought the reader might want to compare the energy of a hurricane to some standard that they may be aware of. I believe each reader would have an idea about the atomic bomb as a reference. So, here goes. The energy released in the atomic bombs dropped on Hiroshima and Nagasaki during World War II averaged approximately 15 kilotons of TNT, or = 6.3^{13} joules. This equals 5.44×10^{18} joules per day. This is about one-tenth the energy of the hurricane per day. In other words, it would take about 2.4 hours for the hurricane's energy to equal the atomic bomb's energy. The hurricane is nearly equal to ten atomic bomb explosions of energy in a day. When you realize that the hurricane produces this energy over many days, it puts the hurricane's energy equivalent to 100 atomic bombs of 15 kilotons of magnitude if the hurricane lasts for ten days at full strength. Doesn't this make Mother Nature look awesome? Meanwhile, the flooding that exists after the hurricane has passed is energy that is persisting after the hurricane has passed.

The National Hurricane Center notes that a hurricane releases heat energy at a rate of 50 to 200 trillion watts. That is equal to 50^{12} to 200^{12} watts. This is the equivalent of a 10-megaton nuclear bomb exploding about every twenty minutes.[59] While a bomb goes off in a fairly restricted area, causing terrible damage in that area, the hurricanes spread their

---

[59] sec.noaa.gov/primer/primer.html

energy over a very large area. It might not appear so devastating because it is spread out in area and in time.

There is some good news. Although their impact on human populations can be devastating, tropical hurricanes or cyclones can also relieve drought conditions. They carry heat and energy away from the tropics and transport it toward temperate latitudes, making them an important part of the global atmospheric circulation mechanism. As a result, tropical cyclones and hurricanes help to maintain equilibrium in the earth's troposphere and maintain a relatively stable and warm temperature worldwide.

These systems are also the most efficient movers of vast freshwater supplies to parts of the world. Thus, hurricanes and other weather systems of this sort provide freshwater to the world's far corners. The winds tend to spread out the heat energy to the parts of the world in the temperate zones.

As indicated, the heating of the water, especially the surface water, causes the air above it to expand, lowering the air pressure. The combination of lower air pressure and elevated temperature of the water (24 degrees Celsius or 75 degrees Fahrenheit) results in a dramatic increase in evaporation from the surface of the water. The normal rotation of Earth causes winds and starts a motion from the West toward the East of the clouds formed by the pickup of the evaporated water. These higher winds increase the evaporation rate. As this happens, large amounts of evaporated water rise rapidly to the troposphere. Here, the heat is released into the atmosphere through the condensation process. This provides the energy that fuels the whole system. This may result in the initiation of several storms. These storms may group together if the water is warm enough (24 degrees Celsius or 75 degrees Fahrenheit) and begin to spin around as one large system.[60] The winds push this large system further across the ocean from East to West. It draws in more warm moist air while gathering energy all the time. This large system begins to spin in ever-tighter circles with Earth's rotation causing an increase in speed. When this occurs, the group of thunderstorms, or

---

[60] aoml.noaa.gov/hrd/tcfaq/D1.htm

the one large system, will develop into tropical storms. The major areas where hurricanes are born are located in the Indian Ocean, the southern portion of the Atlantic Ocean just west of Africa, the Caribbean, and Southeast Asia. The hurricanes striking the United States are initiated off the western coast of Africa and the Caribbean.

The center portion of the hurricane, the eye, is a calm area of low pressure. The eye is relatively calm with winds of 15 miles an hour (25 kilometers per hour). The eye is initially up to 200 miles (300 kilometers) across. As the winds increase, the air is sucked into the eye. This is the development of the eye. As a result, a 200-mile cross section reduces to approximately 30 miles (50 kilometers) across. The outer edges of the hurricane are approximately 200 to 300 miles in the cross section. The speeds can reach 220 miles per hour (360 kilometers per hour) on these outer edges.

The hurricane moves along at approximately 75 miles per hour (123 kilometers per hour). With this rather slow progression, a hurricane crossing the Atlantic Ocean can be sighted well before hitting the islands of the Caribbean and the coast of Florida. Because of the slow progression, it also lasts longer as it passes over land. Depending on the storm's strength, the damage can be horrific. The winds, depending on the category of the storm, are shown in the following table:[61]

Category	Damage	Winds (MPH)	Storm Surge (ft.)
1	**Minimal** • Lowest roads flood • Some roof damage	74–95	4–5

[61] en.wikipedia.org/wiki/Tropical_cyclone; Atlantic Oceanographic and Meteorological Laboratory, Hurricane Research Division. "Frequently Asked Questions: Why don't we try to destroy tropical cyclones by nuking them? NOAA; National Oceanic & Atmospheric Administration," August 2001; NOAA, "NOAA Question of the Month: How much energy does a hurricane release?"

2	**Moderate** • Building damage • Mobile homes damaged • Low-lying flooding two to four hours before the eye arrives	96–110	6–8
3	**Extensive** • Mobile homes and some structural damage to buildings • Flooding on coast anything less than five feet above sea level, up to eight miles inland	111–130	9–12
4	**Extreme** • Roofs off houses • Flooding of terrain lower than ten feet above sea level will flood • Extreme damage to coastal buildings	131–155	13–18
5	**Catastrophic** • Some buildings blown away • Mobile homes gone • Beachside buildings gone • Lower fifteen feet above sea level will flood	156	18+

Table 5. Ranking of hurricanes.

Most of the damage is due to flooding with storm surge up to 20 feet high. Most homes are evacuated up to levels higher than 20 feet above sea level.

Feeder bands of clouds streaking out from the extreme edges of a hurricane are not to be ignored. They pull up moisture from the ocean, feeding the storm's clouds. A day or two after the hurricane has passed, the remainder of the feeder bands will typically come ashore, dropping incredible amounts of rain. The flooding caused by these rains will often cause more damage than the actual hurricane.

New Orleans and Hurricane Katrina are perfect examples of how a horrific storm could cause major damage to property and peoples' lives. Mainly due to the water damage as the hurricane forced water over the levees, Hurricane Katrina forced the evacuation of the entire city of New Orleans.

Hurricanes lose their energy when they pass over land. The cool land cannot support the warm water needed to fuel the energy for the hurricane. Hurricanes can also lose their power if they move too slowly. As they pull the warm water (over 80 degrees Fahrenheit or 26 degrees Celsius) down to a depth of 150 feet (50 meters) from the ocean, the water left behind is cooled. If the hurricane does not move fast enough, the warm ocean water fueling its energy is cooled to the point that the hurricane loses its source of energy and collapses.

This same type of action might occur if there is a tendency for global warming. If there is a tendency to increase evaporation from the oceans of the world by increased heat from global warming, it might have to impact this heating rapidly. A rapid movement of the water being evaporated would follow. If not, the water being evaporated is in the way of the sun's rays, preventing further heating of the water. In essence, the first evaporation provides a shadow over where the following evaporation would normally occur. For example, heavy evaporation may result in water vapor not moving from the point of evaporation quick enough to allow further heating of the water to allow the runaway situation to occur.

At this point in time, it is not obvious how our planet would react to global warming. We can theoretically determine what would happen, but many variables are in this system. The oceans provide a huge mass that has to be overcome to make any significant changes. I have a hard time seeing any major effects on hurricanes. A one degree Fahrenheit change in temperature might not constitute enough of an energy change relative to the 75 degrees Fahrenheit required to start a hurricane.

PHASE TWO
Energy Balance and Man

We have now reviewed how Earth looked just before man entered the picture and essentially how it has looked since. Using the Sun's constant daily energy as the forcing function, we have seen how there were several balances. One was between the Sun's energy and that energy being reradiated back toward space. Another balance was between the Sun's energy and its balance with the world's water systems to supply a daily amount of fresh water. Another balance was with the weather and its annual amount of precipitation. So far, we have not seen any impact by man on these balances. With man around, these balances have continued to exist. What impact does man have on Earth? Keeping all this constant and reviewing man's role on Earth, we will determine his impact, if any, on Earth's environment.

Until now, almost everything discussed related to energy and transfer of energy. Initially, the energy comes from the Sun. Then, the evaporation of water in the oceans transfers the energy to the clouds. Then there is the precipitation that transfers energy to the atmosphere above and out into space. There are ocean currents that spread the Sun's energy. There is energy of the weather's winds, which I call the "great mixer."

It mixes and spreads the joy or harm around the world. There is the generation of oxygen, water vapor, and carbon dioxide and the working of photosynthesis. There are plants and rather simple forms of life. There are volcanic eruptions that generate water vapor, huge amounts of carbon dioxide and many different gases however; this is not a direct result of the sun's energy. This is more related to the energy that earth possessed from its birth and the early years of differentiation and energy brought to Earth via meteors. It is also related to the amount of energy being released by radioactive material in the earth's mantle. These are all energies that happen as part of Mother Nature. Generally, they are fairly efficient and have nothing to do with man.

Man and Other Animals

The amount of energy supplied to Earth by the Sun on a daily basis is equal to 1.5×10^{22} watts per day or 3.6×10^{18} Calories per day. Approximately 23 percent, or 8.28×10^{17} Calories, of the Sun's energy is directed to the water cycle that provides freshwater to the world each day. Approximately 1 percent, or 3.6×10^{16} Calories, of the Sun's energy is directed toward the wind and ocean currents. Only 0.023 percent, or 9.36×10^{15} Calories, of the Sun's energy is directed toward photosynthesis. Forty-two percent, or 1.512×10^{18} Calories, is directed toward warming the land and water on land to the comfort level we enjoy. The remaining energy is reflected back toward space from the clouds, water, ice, and Earth.[62]

I assume that man's entry onto Earth would only have impacted the 42 percent, or 1.512×10^{18} Calories, of the Sun's energy that is used to heat the land and water on land each day. This assumption is based on the fact that this is the only energy from the Sun that is not dedicated to the functions as mentioned. So, during this phase of the book, I will be analyzing how much energy man or his activities consume. I want to see if they approach the energy partitioned off by the 42 percent of

---

[62] En.wikipedia.org/wiki/World_energy_resources_and _consumption; Energy Information Administration, United States Department of Energy, "World Consumption of Primary Energy by Energy Type and Selected Country Groups, 1980–2004," July 31, 2006.

the Sun's daily energy. If man's energy or his activities approach this amount, then there is a major problem.

Man, as we know him, has been around for several thousand years. Around 300 BC, only 100 million humans were on Earth. Now, 6.5 billion are here. Each has a small impact on the Earth and perhaps its climate. But, with 6.5 billion, the effect can be large.

In addition to man, an infrastructure is required to support man. This includes all the animals and plants needed to keep man alive. When you look at the food crops and the number of cattle, pigs, chickens, turkeys, and other animals on earth, it is a large number. Each of these animals takes in oxygen and expels carbon dioxide. Let's look at how this number of people and this infrastructure can impact the balance of the constant energy that our Sun provides and if global warming can emanate from this inclusion of man.

Let's look at one human first. Let's understand his needs and what those needs entail. A human has an average temperature of 98.6 degrees Fahrenheit (37 degrees Celsius) and must maintain that temperature within plus or minus 1 degree Fahrenheit (approximately 0.55 degrees Celsius). If this average temperature falls outside of this range, he suffers from a sickness. Everything that man does while sleeping or awake requires energy. In order to survive, he must take in nourishment in the form of food to replenish the amount of energy he burns off each day. Assuming we are discussing a mature adult, the amount of nourishment required to maintain a given body weight is approximately 2,000 Calories for a man and 1,500 Calories for a woman.[63] A calorie is a unit of energy. There are small calories that are equal to 4.184 joules,[64] the international units of energy. In the United States, we use watts which are equal to joules per second in place of the joule. So, a small calorie is equal to 4.184 watts (joules per second).

The more familiar term used in most of the world is the Calorie, which we use to measure the energy of table foods. This Calorie is called

[63] Some papers written on the subject claim it is 2,500 calories for man and 2,000 calories for a woman.
[64] One joule is equivalent to one watt per second.

74

the large calorie. It is equal to 1,000 small calories, or a kilocalorie. The Calories that a person needs each day are these large Calories. Since it is equal to 1,000 small calories, it is equal to 4.184 watts (small calorie) x 1,000 or 4,184 watts per large Calorie. This large Calorie is the energy required to raise 1 kilogram (2.2 pounds) of water by 1 degree Celsius.

I have used this large Calorie in all the comments made about Calories in this book up to this point. I will continue to use the table Calorie in all of my discussions. This will be recognized by the large C in Calories. I will use Calories in place of watts in discussions of power. Keep in mind that watts are joules per second, so the Calories being used relate to the time of one second. I will convert the Sun's energy we receive each day to Calories. With these Calories being used throughout, it will be easy for the average person to relate to this measure of energy. It relates to those of you who go on a diet to lose or gain Calories. When I write about the Sun's energy in Calories, you can relate to the food you eat and the energy a human dissipates each day. Now let's examine how that plays out in energy used by the average male or female and what effect, if any, this has on the total incoming energy from the Sun on a daily basis, or, more specifically, to the 42 percent that is used to heat the land and water on land.

Let's assume the average male burns off 2,000 large Calories of energy in a day. This is equal to 2,000 x 4,184 watts per large Calorie. As a result, he burns off 8.4 million watts per day. This is listed as 8.4 megawatts, or 8.4×10^6 watts. This is approximately the amount of energy a man burns up during a day while breathing, walking, running, working, and engages in other activities. It is higher in the United States. It is estimated to be between 3,000 and 3,500 Calories per day. I will use the smaller amount because it relates more to the rest of the world's use of daily Calorie intake. Besides, a person in the United States shouldn't be consuming that amount of Calories per day. That is why many Americans are overweight.

If he eats 2,000 Calories during the day, his weight will remain constant. If he eats more than 2,000 calories, he will gain weight unless he burns off those excess calories. If he eats less than 2,000 Calories in a day, on average, he will lose weight. He must make sure he uses up those 2,000 calories through his normal body functions and activity level. If he does not, he will gain weight.[65] This is why one is told to lose weight you have to take in less Calories and do some daily exercise. If a man reduces his daily intake of Calories by 500 a day, he will lose one pound in a week, assuming he continues his normal activities or he exercises.

It is interesting to compare this to a common element used every day all over the world. Let's look at how a male equates against the energy burned by a 100-watt light bulb. Remember I said that 2,000 calories a day is equal to 8.4 million watts per second. This is equal to 2,333 watts per hour. Since there are twenty-four hours in a day, it is equal to 97.2 watts per day. So, this man will burn off about the same power as a 100-watt light bulb that stays on twenty-four hours a day. If the light bulb stays on for eight hours a day, the man will consume about three times the amount of energy burned off by that 100-watt light bulb on that day. If a man in the United States consumes and burns off 3,000 Calories per day, it is equivalent to a 150-watt light bulb in this example. It is also equal to a one horsepower motor running for a little over three hours.

A female burns about three-fourths the energy of a man, so she will burn off energy equivalent to a 75-watt light bulb that burns for a full day. If the light bulb is on for only eight hours of the day, she will burn off thee times what that light bulb would burn off. The American woman consumes approximately 2,500 Calories a day. Her energy consumption must be raised accordingly to offset this larger amount of Calories than other women of the world consume. To lose weight, the woman must take in fewer Calories than she consumes. A reduction of 500 Calories per day for a week is required to loss one pound. So, if a person is consuming 2000 Calories a day and wants to lose weight he/she must reduce the intake to 1500 Calories per day.

These examples of how many Calories a male or female person burns in a day should provide one with an idea of energy consumption.

---

[65] To lose a pound, a man must burn off 3,500 calories. If he takes in only 1,500 calories for seven days, he will lose a pound in a week.

Remember, we are trying to look at the energy that the Sun puts on Earth each day. We are trying to determine if man's entry into the picture causes any problems with energy since we know that global warming requires additional energy.

Let's look at the total population of 6.5 billion people. Let's assume each burns an average of 1,500 Calories per day for a total of 0.97×10^{13} total Calories. For a round number, let's decide that man burns off 1.0×10^{13} Calories per day. If we want to express this in watts, we know that one Calorie equals 4,1868 watts. So, we multiply the Calories (1 $\times 10^{13}$) $\times$ 4,1868 watts per Calorie for a total of 4.187×10^{16} watts each day. This is a lot of energy. The Sun supplies 3.58×10^{18} Calories per day, or 1.5×10^{22} watts per day; much of it heats the oceans and provides evaporation so man can have freshwater to drink. Man makes a large demand on this system of energy. Remember I mentioned that the land takes 1.512×10^{18} Calories per day. So far, we have identified that six and a half billion men/women burn off 1.0×10^{13} Calories per day. Now we have to determine how much energy the infrastructure burns up to supply man's daily needs. I could determine the total required counting delivery trucks etc. but I am only going to count the ones where Calories are supplied for food.

Man's Food Source (And all the other Calories that are needed to support man)

Where do these people get this amount of energy per day? Most of the energy comes from the sun. It's no different for man. He gets his from the sun, but it's an indirect source. He consumes food that was derived from the Sun's energy. The food is in the form of plants, animals that eat plants, and their derivatives that the person eats each day. The plant is the basic staple in this listing. The plant derives its energy from the Sun. Of the 6.5 billion people on Earth, approximately 4 billion rely on plants for their food; with the other 2.5 billion receiving most of their daily requirements from meat in one form or another. So, each day, humans must depend on each person receiving approximately 1,500

Calories (I will use this conservative amount) per day. Multiply 6.5 x 10^9 people on earth by 1,500 Calories per person. That is a total of 9.75 x 10^{12} Calories per day. Let's round this number to 10^{13} Calories per day for the world of people.[66] That's a conservative number of Calories per day. That's 10,000 billion Calories per day (or 10 Trillion). Let's review how we come about these Calories.

To relate the capacity of the energy required in the United States for plants and animals (not including man), approximately half of the land in the United States is required, along with approximately 75 to 80 percent of the freshwater and about 20 percent of the fossil fuel energy consumed in the United States each year. This is an enormous amount of land and energy required plus the huge amount of a scarcity such as water. But, this type of energy and resources must be directed to feed people and the animals that are then consumed for food and other uses. Of course, Americans consume more like 3,500 Calories per day and supply an enormous amount to the other people in the world who are not as fortunate to have the natural resources of the United States.

Plant Calories

Plants must grow for a period to be edible. Let's say four months is a good estimate. During those four months, they received energy from the sun and soil. They also received carbon dioxide and water that was required to complete the photosynthesis process. Then the plants were harvested. Man used his energy or the energy of machines. After harvesting, the plants were transported to the market. At the market, people expended their energy to place the produce or meat out for the patrons to see and buy. After that, the patrons transported the food to homes, where they were stored in a refrigerator or eaten promptly.

In many cases, the food required energy to cook the food so that it would be edible. The person eating the food expended energy to perform this function. At any point in time, this is now happening all

---

[66] I am only using 1,500 calories as the average per day.

over the world to provide each of us with food for the day. The next day, most of those actions are repeated.

If it were not for the Sun, none of this would be possible. Let's assume that half of man's daily requirements of Calories come from plants. That would be 0.5×10^{13} Calories per day. Let's assume it takes at least that amount of calories to bring these to your table for consumption. On a daily basis, the plants' supply must equal 1.0×10^{13} Calories. This extra half is probably understated when you consider that, for every plant being eaten, many are not being picked and are receiving the Calories of the Sun and photosynthesis on that day. Consider the energy used for fertilizers, water, and fuel to make it to man's table. Other than plants, let's look at where the Calories come from.

To provide some insight to this amount of energy and where the resources come from, meat-based food requires more energy, land, and water than the vegetarian diet. The large number of food animals in the United States uses a large amount of land. Nine billion animals, including cattle, cows, pigs, and chickens, are being maintained in the United States to supply the meat requirements that are consumed each year for food.[67]

Whatever meat Calories are eaten, there are eight times as many that are in the food chain, taking on Calories and being fattened up for future food. The Calories that these animals are consuming must be counted each day. They are in the queue to ensure there is food in the supply chain that is available for food each day. When the Calories that are consumed that use meat as the food, you must multiply that figure by eight to show the Calories being consumed on a daily basis by these animals to provide man's future consumption. This is not one of the places where "just in time" can be used. We cannot grow them and consume them in the same day or year. Most cannot reach that stage in a year.

In order to supply the meat for man on a daily basis, we must have approximately eight times as many animals in the energy queue to supply the daily requirements.

---

[67] upc-online.org/slaughter/2000slaughter_stats.html

Meat

In order to obtain meat for some of our calories, we must depend on raising the various forms of animals that are eatable. Anything that is not a plant falls in this category.

Cattle

Each year, 1.3 billion cattle, out of the total population of 9 billion, provide food for the given year; some of which become beef, milk, cheese, or some other form of food Calories.[68] These cattle derive their energy from the vegetation they eat. Again, we can thank the sun. Of course, the vegetation comes in many forms for these animals, but most is through grazing of some form of grasses. Farmers also provide hay, wheat, corn, foliage, and other nutrients. Some critics say it might be a better use of the food to provide it directly to man rather than through the indirect chain of the animals. However, many nutrients are supplied through the cattle chain that could not be supplied directly to humans. For example, man cannot digest cellulose and certain other carbohydrates that cattle are able to do because of their four stomachs and the ability to chew cud[69] In addition, man cannot take nitrogen directly into his system. It must come from the digestion of plants or animals that eat plants. Nitrogen and phosphorus are needed for some of the amino acids and other biological requirements for man. While supplying many Calories for man's consumption, cattle require a considerable amount of Calories each day to exist. This is the use of some of the sun's provided energy. Meanwhile, cattle generate a considerable amount of methane mainly through burping. This methane is a greenhouse gas that we must consider when reviewing the sources of global warming.

[68] en.wikipedia.org/wiki/Aquaculture; J. Hepburn, "Taking Aquaculture Seriously," Organic Farming (Winter 2002); M.B. Timmons and others, *Recirculating Aquaculture Systems*, 2nd ed. (2002).

[69] Cud is a form of food that cows taken in one of their stomachs so they can eat again, chew, and eventually digest.

Each of the 1.3 billion cattle requires approximately fifteen times the daily Calories of a human. So, the Calories add up to $1.3 \times 10^9 \times 30,000$ Calories for a total of 3.9×10^{13} Calories per day; resulting in other food products besides beef. For example; milk, cheese, and hides. Interesting enough, cattle as well as pigs also provide insulin for people that have sugar diabetes.

Chickens

Each year, approximately 60 billion chickens are consumed for the production of food.[70] At any given time, about 24 billion chickens are alive and well. Chickens provide two main sources of food: their eggs and their bodies. Astoundingly, there are four chickens to every human. It is also astounding to consider the amount of food they provide. In some ways chickens are better food for man than what cattle provide. They have less fat, so they allow for fairly fat-free diets. I do not know what the world would do without eggs for breakfast or many supper meals. In addition, chickens require much less food and care than cattle do. Many poor countries can maintain chickens because they take up less space and require less food supplied daily. Chickens can survive on rather poor land. Many of the sources of chicken are raised in small pens or cages, thus allowing for the efficient supply of food.

The biggest countries in the production of chickens for food sources are: United States (9 billion per year), Brazil (7 billion per year), and China (5 billion per year). These are estimated numbers, and they add up to 21 billion of the 60 billion a year consumed in one way or another. The other countries of the world provide the other 39 billion. (49)

Almost 15 million pounds of other poultry (Turkey, Pheasant, small wild birds, Quail) was slaughtered for food. Chickens consume ground corn, weed seeds, and many other foods of this nature. The chicken is highly active and burns calories rather profusely, especially the hen during her egg-laying sessions. Generally, chickens are eaten

[70] en.wikipedia.org/wiki/Mount_St._Helens; D.R. Mullineaux and D.R. Crandell, "The Eruptive History of Mount St. Helens," USGS Professional Paper 1250, 1981.

completely from the top of their highest part of their body to the lowest part. Chicken eggs are eaten in many different forms. They are also used to culture flu vaccine.

A chicken's daily intake of Calories is approximately 2,890 per day. The total intake of calories per day worldwide is therefore $60 \times 10^9 \times 2890$, which equals 1.734×10^{13}. Despite their size, they provide the most food value for the world intake.

Pigs

In terms of global production, pork is the number one meat, accounting for more than 40 percent of all meat produced. The highest consumption is in Europe, where approximately 110 pounds (50 kilograms) is consumed by each person in a year. That is about twice the amount consumed by each person in China and the United States. Europeans enjoy such products like bacon, ham, pork chops, pork roast, and sausages. (50)

The pig is a favorite because it does not take up much space. In most cases, it eats select garbage of humans. Many farmers use their crops for food and the leftover crops for the pigs. Proper smoking of the slaughtered pig results in a long shelf life, so the food source can be spread out over time more efficiently than other forms of meat. There are approximately 2.4×10^9 pigs worldwide. Besides their use as a food, their skin is used to make leather for various sporting events as well as other garments for man.

The Calorie intake of a pig depends on the size of the pig before it is slaughtered. Pigs fall into two main weight groups: those that are approximately 11 to 66 pounds (5 to 30 kilograms) and those that are 66 to 220 pounds (30 to 100 kilograms). The smaller of these takes in 3,000 Calories per day; the large ones take in 8,000 Calories per day. Using 5,000 as the average number for the total of 2.4×10^9 pigs; this would be a daily intake of 1.2×10^{13} Calories per day worldwide.

Fish

In 2003, the total world production of fisheries product was 132.2 million ton or approximately 2.64×10^{11} pounds.[71] This amounts to approximately 40 pounds per person in a year derived from the fisheries. This is an understatement for many of the countries of the world where fish is the main source of food and protein. This amount from fisheries is approximately one-third of the total amount of fish consumed. Two-thirds of the fish consumed are caught wild. The percentage from the fisheries has been growing as many new fisheries are started each year. All in all, this adds up to 8.0×10^{11} pounds of fish consumed, wild or farmed, per year.

Efficiencies of fisheries have increased tremendously. At one time, the food provided to each fish was two fish for each fish produced. This has been improved to one fish per fish produced. This may sound inefficient, but the one fish supplied is not high on the food chain for man's consumption. Essentially, one fish that man cannot eat is supplied to provide a fish that is edible. Food for fish is also supplied in the form of ground kelp or other ocean food.

The source of wild fish has remained essentially constant over the past decade. This number must be doubled to account for the food the fish consumes of one fish per fish in the fish farms. In the wild, the numbers are about the same amount. This means that 1.6×10^{12} pounds of fish are consumed to provide food for the fish that will go toward man's eventual consumption. However, at least ten times that amount of fish are free and consuming food each day. So, the calories consumed by the fish eaten or the fish being maintained represent at least 1.6×10^{13} pounds of fish per year. For 365 days per year and approximately 365 calories per day consumed, that equals approximately 1.6×10^{13} calories per day that man consumes. This is probably well understated to represent the amount of calories actually being consumed or swimming around in the wild. But I chose to use this number because it is relatively consistent with the other sources of food I reviewed. The

---

[71] math.ucr.edu/home/baez/temperature/

fish is the hardest for which to make an accurate estimate. If we knew where they all were, we could catch more of them each day. China now produces percent of the world's farmed fish.

In the United States, approximately 90 percent of all shrimp consumed is farmed and imported. In recent years, salmon aquaculture has become a major export of Chile, especially in Puerto Montt and Quellon.

Summary

Obviously, a significant portion of the world's energies are directed to providing the 1,500 to 2,000 Calories per day for man. In doing so, the providers expend many Calories as they work to bring the food to others as well as themselves. We cannot count their individual Calories because they are already included in the Calories determined for the world's population. However, any energy that machinery provides that provides these foods must be counted. In addition to the Calories to keep these animals alive, clean water is required for all except the saltwater fish. In most cases, this water must be transported or pumped. Even the saltwater fish in fisheries must have their food and water transported. Note that most of the calories from meat sources derive their energies via the consumption of plants rather than other animals. Of course, the energy of these plants was derived from the sun. Consider all the fuel consumed to provide these foods for man. This is a terrific number of Calories.

To simplify the summary of calories required to keep people alive, let's assume that all of their daily Calories are derived directly by eating the plants or indirectly by eating the animals that ate the plants. This means the Calories derived from the Sun to raise the plants are only counted once with respect to providing man's daily requirement. However, the calories to keep the animals alive before man consumes them represent table Calories on a grand scale to support man. Let's add them up in the following table:

Food	Amount of Energy (Table Calories) Required to Support Man
Cattle	3.9×10^{13}
Chickens	1.734×10^{13}
Pigs	1.2×10^{13}
Fish	1.6×10^{13}
Plants	1.3×10^{13}
Sum Total	**9.74×10^{13}**

Table 6 Amount of energy (table calories) required to support man.

This number (9.74×10^{13}) represents Calories just in the food value for man to replace the Calories he burns off. Because this was on the conservative side, let's assume 1.0×10^{14} Calories. That number is approximately ten times the daily requirement of the world's population. That's probably a reasonable approximation because it takes many of these plants and animals to be in the loop to be able to supply man's daily requirements.

This daily amount of 1.0×10^{14} Calories is compared to the Sun's energy supplied to warm the land and water on the land of 1.512×10^{18} Calories per day. So man's consumption of food requires 0.00661 percent of the Sun's total energy that provides Calories to Earth's land and water on that land. At present, the animals used for food consume approximately eight times as much grain as man consumes in the United States.

This amount of food energy does not take into account the huge amount of Calories taken in each day via the use of sugar as a sweetener. Think about the number of soft drinks each person drinks a day and the number of candies eaten. All of these just add on to the calories to support man, but these are taken in as entertainment rather than as a basic need. As a result of this additional consumption, man takes in more than 3,000 Calories per day. This increases the support needed worldwide to maintain this additional weight. I was considering finding the amount of soft drinks that are sold per year and calculating this total Calorie intake, but I decided to just state that it is a high number. We can just add it to the total Calories consumed worldwide in a day or a year. Since part of the soft drinks comes from plants, including the sugar, flavoring and other ingredients, they are accounted for in some of the details on plants. However, these drinks do carry a considerable amount of Calories and for people that want to lose weight, here is one big place to do it. I realize that many of the drinks are sugar free now and called by various names to indicate they are low on Calories.

Keep in mind that I have not included the Calories burned up in transporting these foods over many miles to their place of consumption. Also consider that I didn't include the energy for cooking the foods. Some foods come in cans and therefore are transported to one place for cooking, placed in cans that require energy in the canning process and then are transported to the grocery stores and eventually into the home for consumption. Lots of energy involved here.

Why go through all this Calorie counting?

Why go through this calculation of Calories? Remember, we are trying to determine if there is global warming. In order to have global warming, one must find the energy to provide that extra warming. Therefore, we review the various energies being used each day and determine if something is done inefficiently to provide extra energy for global warming. If we do not find the energy through this method, then we must find where new energy is coming from to provide global warming. Because there is a large infrastructure to supply man, I have reviewed it to see if we can find any clues. This is to show part of the infrastructure required to be maintained in order to feed man. It is also an effort to determine how much of the Sun's energy is needed to support man in addition to all the other things that have been discussed about the Sun's energy and where it is used without man. When you consider that the main use of the Sun's daily energy is to supply energy to the oceans to provide fresh water for man and almost all of the rest of it is used to handle the requirements for man's food, it provides a picture of a world made for man. It also may indicate some new problems to be concerned about. You will see some of the surprises that come out of this study. It is also a study to determine when the Sun's energy can no longer sustain any growth of man and animals. With the population growth expected over the next twenty; to thirty years, there could be major problems. We shall see.

Animals all breathe in oxygen and breathe out carbon dioxide. Taken as a whole, they put out approximately six times the carbon dioxide of man. In addition, the cattle and pigs put out a considerable amount of methane, a greenhouse gas. I am interested in comparing these needs of man versus the problem of satisfying the basic requirements to keep global warming from being the major problem.

Interestingly, this number of Calories to support man is all supplied by the energy of the Sun, if you ignore the energy derived from fossil fuels to provide motion to vehicles that deliver the food. However, man derives no direct energy from the Sun. All of the Sun's energy is used to provide photosynthesis to provide the growth of plants for man to eat or feed the livestock of the world for man to eat. Meanwhile the Sun's energy provides the water cycle that provides man the freshwater to drink and energy to keep Earth warm. Man could live in a structure without any Sun; if he had these foods brought in on a daily basis, he would live a normal life.

Energy Flow

In order to better understand the means by which Earth receives and gives up energy, you must understand certain terms that are used when discussing energy. At this point, I believe it is important to understand some recent history of Earth, that is, history before man and animals appeared on the scene.

Energy is the only thing that moves on Earth. There are other forms of energy without movement, but they are relatively stagnant and exist as potential energy. But there is no release of this potential energy without movement. This movement takes many forms and takes on energy through the kinetic activity of the body's atoms or molecules by taking on or losing heat energy. Movement, or the loss of heat, is the only way of losing energy. But it is not really lost. Heat emanating from a man or animal or any body is supplied to the atmosphere around these heat radiators. This heat energy can escape to space, or be transferred to other bodies. This movement may only be at the atomic level, and the observer may not notice it. All movement results in the loss of energy either in the mover or the moved, no matter what form the original energy was. However, the total energy, including the mover's loss or the moved's gain, plus any loss of heat energy or emanating sounds,[72] equals the total initial energy. Energy cannot be destroyed. It can only change its form. Generally, you have to figure out where the heat went.

Without people, the energy in the world is very efficient. The Sun, oceans, water cycle, precipitation, photosynthesis, and water falling from great heights do lose some transfer of energy to noise, but they all go about their business very efficiently.

Sun is the big loser. It casts its light upon Earth, transfers it, and loses some of its enormous energy constantly. The Sun is gaining entropy (loss of energy from a closed system is termed a gain in entropy) and losing energy. It is lost to the sun and gained partially by Earth but much of Sun's energy goes out into space, where it is not lost but used somewhere in space.

We can assume that the only energy entering Earth is the energy that the Sun provides. Other entries, such as small meteorites, result in a minor change to Earth's potential energy. Most burn up in the upper atmosphere and do not contribute to the gain as the heat goes off in space. We, therefore, can consider the Earth and the atmosphere up to a couple hundred miles just outside Earth to be a closed system. The sun's energy coming in and what is lost by radiation from Earth out of this system is in balance. We have been considering this to be true for many years. However, even though we can measure the amount of energy that the Sun supplies very accurately, I do not believe we can

---

[72] A form of energy required for the movement of sound waves

measure the amount radiated out as accurately and determine if it is equal. Needless to say it must be equal or Earth temperature would change.

If they were equal, there would not be any energy to support the growth of man on a daily basis. Or would there be? Where would our daily source of energy come from? Would we just be using the energy that was always on Earth? Are we recovering it and using it up?

Plants would not grow without sunlight, so they must use some energy that is not returned to space to balance the amount of energy coming in from the Sun. The Sun's energy, carbon dioxide, and water vapor provide photosynthesis, which allows the plants to grow and oxygen to be supplied to the increasing number of animals and humans on Earth. It is true that, as humans die, they become unusable energy for many years. Eventually, their remains may supply some useable energy. But what about all the inefficiencies in the world that result in loss of useable energy? Or is it still useable? These are tough questions to answer. We know it is not lost.

If Earth's energy is in balance with the energy that the Sun supplies, then we have to determine how that happens. The same energy was being supplied by the sun since Earth was a couple of million years old, and there were no animals or plants on earth. We can assume that man is part of the determining equation. We can also assume that man does not take in any of the Sun's energy directly, except for that sunburn and Vitamin D. The energy is given directly to the plants. The plants are a wonder of nature because they do not have to eat other things to gain Calories. They use sunlight, carbon dioxide, water vapor, and photosynthesis to take on Calories. Man and other animals eat the plants to gain this energy. In addition, man eats the animals that have eaten plants or other animals to provide him his daily quota. This also supplies man with the nitrogen he requires. Plants are wonderful when you consider that you can take a seed from the plant and grow another plant. In most cases, you can take many seeds and grow many of the plant. In this way, the plant gives away some of its earned Calories to support this circle of energy.

Man's energy is given back to space via the 2,000 Calories per day that he consumes to help balance the indirect heat intake from the Sun. This energy does not just disappear. It is emitted as heat in the form of infrared wavelengths, water vapor from his sweat, and the moisture he breathes out. Carbon dioxide is expelled to provide the one necessary ingredient for plants to grow. Energy transfer occurs when he causes something to move. Man's blood delivers oxygen to all the cells in the body. In order to accomplish this task the oxygen that man breathes goes to his lungs. In the lungs there is blood that is being returned from making its flow around the body. This blood is returned through the veins, and is depleted of much of its oxygen. This blood has a look of a different color than most of the blood one sees. It has a bluish cast to it. This blood contains iron which has been depleted of its oxygen. This iron is capable of taking on oxygen in the lungs. The oxygen chemically reacts with the iron and becomes hemoglobin (one heme atom consists of an iron atom and one molecule of oxygen) and the blood takes on a rich red color. The next heart beat and this oxygenated blood is pumped from the lungs to the part of the heart that now pumps this rich oxygenated blood to the body. This is done via the arteries flowing from the heart to all parts of the body. The arteries from the heart are big in diameter and as they flow through the body they spread out through small arteries called capillaries. As the oxygen is delivered and taken up in the parts of the body the blood starts its return toward the lungs, first in small capillaries and then into larger veins called the superior vena cava which is located around the neck area. This oxygen low blood is returned to the heart by the vena cava and enters the heart in a different chamber from the one it left. This chamber at the next heart beat pumps this oxygen poor blood with its iron still in tact to the lungs to be re-oxygenated. You probably heard of people that have a low energy level and it is said they have "iron poor blood". This relates to the fact that if the iron level in the blood is low, then when it goes to the lungs it cannot take on the amount of oxygen needed to reenergize the cells. So, the person really has two problem, low iron level and low oxygen level. When the blood has a low oxygen level it cannot supply ample oxygen to the cells of the body that provide the energy for the body. The person becomes weak.

Man's everyday existence on earth requires that energy of his body that is used for some task, including staying alive, is replaced. Man's requires energy for every breath he takes. All his organs require energy to carry out their tasks. Man uses his sight to see, which takes energy from the brain. When man is thinking he is using energy from the brain. Any physical exertion requires energy that must be replaced. It is not important that it be replaced at exactly the time it is used but it must be replaced within days of use unless the person wants to lose weight. The same is true with water, only it doesn't contain any Calories, but man must have the water in order that the body retains enough fluid to carry his energy around and supplies fluid for critical function. Man must return all the energy that was used to keep him alive. On an average, every day, he takes in and expels 2,000 Calories in some manner. The same is true of all the animals roaming the fields to supply man with Calories. They burn off energy that is replaced by the plants they eat each day before man consumes them for one of his meals. The animals must emit heat in the form that man does, that is, heat is given off in infrared wavelengths. As a result, they give a lot of energy back into space as infrared wavelengths. Where man must replace his 2000 or so Calories, the animals must replace the greater number of Calories they use on a daily basis. Go back to the chart and see the listing of the various animal kingdoms and see how much they must replace in energy each and every day to stay alive. In many cases man overfeeds the animals to fatten them up so as to provide a greater amount of food per animal. This is a normal routine.

Using this form of argument, man has taken this step to remain atop the food chain. He has improvised since the time he came on earth to ensure he is able to acquire the needed food to keep him alive. Early in the life of man he only had plants to feed upon but then began to eat the meat of animals. As time went on he learned how to domesticate many of these docile animals to ensure a supply of food. The plants and meat consumed each day add up to the Calories he needs to survive and live a useful life. As man consumes these foods and remains active he gives off heat in the form of infrared heat. This heat is provided to the atmosphere around him and becomes part of the heat that escapes to outer space. By this heat being exhausted to outer space it allows earth to remain at a constant temperature, and balancing the heat that is supplied by the Sun each day. In addition, man and the animals provide the perfect balance with the plant world by giving off carbon dioxide that is taken in by the plants. Here we are back at the balance between plants that take in carbon dioxide and give off oxygen and animals that take in oxygen and give off carbon dioxide. Keep in mind that for every human there are many animals that provide his food that are also expelling much more carbon dioxide than man. Its interesting that man breathes in carbon dioxide while breathing in oxygen and this is transferred to the blood and makes it journey around the body but is not used and is expelled during the breathing out of man. He must breathe out the carbon dioxide so that the plants have their supply of their required carbon dioxide.

We can assume that Earth's energy relative to the energy received from the Sun's energy is in balance. Therefore, over the course of a year or some relatively short period, Earth does not obtain any net gain of energy. If in balance, the heat gained from the Sun is equal to the heat energy lost to space; much of it by man and the animals.

However, with a little imagination, one could consider this a case where Earth's entropy shows a gain. A gain in entropy is a gain in unusable energy and a loss of usable energy. This is a questionable point in this discussion. This would be a case where the Sun's energy supplied to Earth is constant. But what is being returned to outer space is not equal to this constant. It would be a reduced rate that Earth radiates to space. The Earth does not gain in usable energy in this scenario, but it gains in unusable energy. It gains in entropy. Because entropy relates to a small portion of energy left after the fact and is unusable in whatever its form, it may go unnoticed and not cause Earth to heat up. It rests on Earth in this unusable form. For example, consider when a car skids to a stop and leaves tire marks on the highway surface. The tire marks are considered as energy that is unusable in this state. If later, there becomes a method of using the material that the mark left, then the energy is recovered. The entropy shows a decrease and this energy is then used or returned to outer space to keep the balance. I keep thinking about this but cannot prove it. Maybe our dead carcasses will lie around for a few hundred million years and provide some future generations with some entropy that becomes useful energy.

Man burns up energy while moving around things that are part of Earth, which are not relative to energy that the Sun provides. It is just part of Earth that he moves around and man dissipates energy while doing this and his dissipation of energy results in heat given off to the atmosphere. When man takes cement from the ground that is part of the Earth, and fabricates concrete to make roads, he is giving up heat while just moving around a part of Earth that is here as the planet is here. He is taking mass energy from Earth and returning it to Earth in a new form. As a result, he is burning off a portion of his 2,000 Calories per day. So, in many cases, man's gained energy from eating plants or animals is used to take energy from Earth and put it into a different form on Earth at the expense of his energy and the gain of some of Earth's energy. For example, if a man lifts a boulder to a higher location, the boulder gains in potential energy at man's loss of energy. So, energy is conserved including the heat given off by man in this exercise. This heat rises and becomes part of the atmosphere to be eventually expelled. Each day around the world, this heat emission happens time after time by the 6.5 billion residents. If man did not eat his 2,000 Calories a day to replace the energy lost, he would lose weight. He soon would not have the energy to lift another boulder to the same height as the original one he moved.

Therefore, as a result of the previous statements, Earth consumes a constant, useable energy. It has maintained this energy over at least a few million years. Man's energy comes from plants, either directly or indirectly. Because the Sun's energy for photosynthesis has been analyzed as being 0.023 percent of the Sun's total energy, or 8.28×10^{14} Calories per day, this amount is barely on the level required to support man's needs, including the animals that are needed to support man's need for food. This would indicate to me that the amount of sunlight required for plants is a variable; as the demand by man and animals goes up the amount of energy from the sun that warms the terrestrial regions of earth must supplement this demand. For example, as the demand for food goes up man must plant more food, and this will shift over more of the proper Sun's ray for photosynthesis. Photosynthesis is a demand type exchange of energy, that is, the more plants there are the more synthesis occurs and the more carbon dioxide is consumed and oxygen generated.

I have reviewed the more or less destructive acts of man when it comes to gaining his energy on a daily basis. However, man is the only creature on earth, except maybe the honey bee, that plants food and plants trees. Plants can't live forever and if man didn't plant the seeds and provide the water than this cycle would only be a destructive one. He not only plants food for his own personal use but for others including the animals that provide him food. Man has been very creative in finding ways to capture that part of the Sun's light energy.

Because energy cannot be created or destroyed, Earth's constant energy must be accounted for over that period. It may change from one state to another, but it may not be destroyed if we consider the closed system that was previously mentioned. The heat given off by man and animals is added to the local environment and raises its temperature. It must be a part of the energy that is returned to space via infrared wavelengths to keep this balance. We have covered how much of the Sun's energy is returned to space as a result of reflection from clouds or from the oceans or Earth, absorption by clouds and then released to space, some reradiation by Earth in the form of infrared, some loss from heat fluxes relative to the heat from the oceans and the resultant weather. Now we have to add this infrared energy from man and the animals that man eats. Some animals are carnivorous and eat other animals that eat plants. In many cases, man does not eat these animals, so the energy goes from one animal to another until his cycle is completed.

For all the calculations I used, I did not use the Calories that man consumes per day. They are all provided by the Sun indirectly to man. Man is a passive recipient of the Sun's energy. Man does use some of those 2,000 Calories per day for energy directed toward Earth's materials and the building of an easier place to live in. Man has a drive toward improving life on Earth. He moves dirt around, builds structures out of Earth's materials, works the land for food, provides means for moving food around, builds equipment for many applications, participates in sports, and participates in reproduction of his kind. Additionally, man gives off energy though heat from his body, carbon dioxide and moisture from his breathing, and perspiration to the atmosphere that will be released to space or recycled in some manner. Man is a significant supplier of carbon dioxide to the atmosphere, but the combination of man and the animals he consumes for food give off a tremendous amount of carbon dioxide; probably more than anything on earth except perhaps the volcanoes . Volcanoes deep in the ocean generate a tremendous amount of carbon dioxide. Man also takes in and gives off a tremendous amount of moisture, but here again; the animals he eats consume about ten times as much water as man. Man and his animals is also significant supplier of heat to the environment. Their combined Calories a day have already been reviewed.

I have not found anything in these activities that results in an increase in global warming as a result of man himself. This is interesting since, as will be reviewed, man and the animals give off more carbon dioxide than the machines of man. Man has been on Earth for a little over 30,000 years. Only recently, has there been an indication of global warming. Man and animals did not give off enough carbon dioxide to impact the climate. This is at least true when the population was at the 1850 levels. In fact, it may be a standoff when carbon is considered. Man eats plants and animals that contain carbon. The plants gained their carbon from man and animals breathing out carbon dioxide. It would seem that, among plants, animals, and man, the carbon keeps making circles by using the energy of the Sun, photosynthesis, and man's consumption of plants and animals that eat the plants. This is the picture I see before the arrival of the industrial revolution and the entrance of oil into the picture, as well as a huge increase of population

However, the review of the energy that man uses and of the energy required to support him brings other things into the picture. This is especially true when one considers the increase in the number of people on Earth. I will display the impact that man has on earth in the section of the book where I discuss Global Warming and more importantly *Beyond Global Warming*

PHASE THREE

. . . .

Enter Oil

One of the most significant contributions to the industrial revolution was the drilling of an oil well in Titusville, Pennsylvania, in 1859. This was the first commercial well drilled for the primary purpose of use in industry. Thus, this began a major source or energy that was separate from the daily energy of the Sun. Oil now entered the picture along with other fossil fuels being used at the time such as coal.

Remember discussing entropy in relation to the meteor that struck Earth 65 million years ago? As a result, unusable energy was stored at the time. In addition, some of Earth's material was left to rot over the years. While under pressure, it provided what is now most of the fossil fuels. The drilling of the oil well in 1859 was followed by a huge growth of oil wells across the United States. They eventually appeared in the Middle East, Russia, and parts of northern South America. More underground energy was put to use with the movement toward central heating that used natural gas. Previously, the main fossil fuel used was coal, but oil could be so easily obtained and transferred, making it the major supplier of earthborn energy during the twentieth century. Oil and its derivatives provided capabilities that no other fossil fuel could provide, mainly it was portable. It was a mobile form of energy and provided more energy in a smaller volume than any of the other forms of energy.

During the twentieth century, there was a major growth in vehicles that consumed gasoline, diesel, and aircraft fuel. The previous century was witness to the major growth of the transportation industry. There are an estimated 650×10^6 (650 million, or one for every 10 person on earth) cars in the world. The amount of energy consumed in the United States on 150×10^6 cars was 3.5×10^7 Calories per day. The rest of the world has 500×10^6 automobiles. Assuming they each use one half of the fuel of a car driven in the United States, on a daily basis their consumption would be 8.33×10^7 Calories per day. So the world consumption is 1.18×10^8 Calories per day. On a yearly basis this consumption is 4.015×10^{10} Calories.

The daily energy consumption worldwide of man per day based on 1500 Calories is 1×10^{13} Calories and for a year is 3.6×10^{15} Calories. I provide this number for a reference to show how the energy consumed by man each day compares to automobiles energy. This is only the energy consumed by man alone and comes from his consumption of food alone.

- This is almost equal to the amount that man consumes per day. This is about 10 percent of the calories required to support a man's Caloric intake. This only considers the gasoline that the average person in the world uses. There is also a tremendous use of gasoline to run tractors and bulldozers. Then there are the airplanes and their fuel requirements, which have a demand that is growing faster than car demands. The mass transit systems consume a significant amount of energy. There is the electricity that fossil fuels generate. There is the home heating and air-conditioning use of fossil fuels to be accounted for. Every act that is consumed in this world requires energy. I did not include these Calories in the support of man in my analysis of the energy consumed by man. These fuels did not come from the Sun's energy. They came from the entropy of the large animals that were killed 65 million years ago when the meteor hit Earth. This was in the form of oil and other fossil fuels, which provided man a major advantage. Now, man has the energy that the Sun supplies on a continuous basis as well as the entropy of the dead dinosaurs that was stored over that length of time. During that time span, the bodies decomposed and went from one type of entropy to another form of entropy. As man uses the energy from something that was once unusable, that something loses its entropy while becoming a useful energy

- The key factor in this critical situation is that most entropy on Earth is the result of the Sun's energy not being efficiently used. It may just be man performing a task inefficiently, resulting in an increase in entropy. In this particular case, the entropy was not from misuse of the Sun's energy. It was from the unusable energy tied up from the crash of that meteor upon Earth 65 million years ago. . The approximate energy released in the formation of the Chicxulub Crater in the Yucatán Peninsula_=1.19 x 10^{20} Calories.

This is extra energy added to Earth from a meteor crash. Now we can put it to use. Isn't Mother Nature wonderful? She saved all this energy over that many years so we could have it available since the middle of the nineteenth century. It is important to keep in mind that all the fossil fuels did not only come from that meteor of 65 million years ago. They came from much of Earth's decomposition of other animals and plant matter that was unused and covered over for at least six million years. Some of it was derived from Sun's energy that was tied up many years ago in some form on Earth. Some fossil fuel such as peat bogs and coal were used as sources of fossil fuel energy much earlier than the oil derivatives.

Global Warming

History has shown that there has been global warming since early in Earth's lifetime. Is there global warming today? There certainly is. We owe our existence to the fact that there is and has been global warming. We owe our present world average temperature of 59 degrees Fahrenheit (15 degrees Centigrade) to the fact we have had and continue to have global warming. This is not my calculation. It is the calculation of scientists and engineers worldwide that have been reviewing and calculating this phenomenon for many years. However, maybe we should not have called it global warming. We should have called it **global warmth** instead. This is a more proper term for the phenomenon since the atmosphere and the temperature we obtained several thousand years ago has been essentially constant; it hasn't been warming. Global warmth is the product of that atmosphere. Calling it global warming connotes a different meaning. It tends to tell a person that a phenomenon is occurring that will continue to increase the temperature of Earth.

How much has the global temperature increased over the past one hundred years? The past fifty years? I do not have firm, firsthand information on the answer to this question. I have read several articles by experts in the field. Some indicate it is about 1 degree Fahrenheit (approximately 0.5 degrees Celsius). Some say it is 1 degree Celsius (almost 2 degrees Fahrenheit). But I have not read an official statement saying that the temperature has increased. If it has changed, the change was so small that it was inconsequential to me at the time of reading the article. It related more to slight changes in the weather or climate due to natural circumstances. I am seventy-five years old, and I have not recognized any major change in the climate. Sure, major things happen every year, like unusual rain in places and high temperatures recorded. I once read a book that said that, each year, at least four records of unusual circumstances will be related to the weather happening around the world. But they normally do not repeat themselves, unless there is a long drought somewhere in the world. Then everyone knows about it. This is especially true with the advent of the increase in the media's ability to communicate these types of events. In other words we hear more about it.

With the advent of the Internet, happenings all over the world are brought to the attention of many people interested in things that may impact their well-being. Scientists who are experts in this field have the answer to this question. There is much disagreement among them as to how much and what caused it. Certainly, it has not increased enough to be concerned about it as a primary function. Neither our president nor any of our representatives have signed the Kyoto Protocol to take any actions to correct the problem, but many nations have signed up to reduce the amount the carbon dioxide generated by our cars and other sources of power. I, along with many people in the world, agree that something should be done to reduce any types of anthropogenic emissions that would impact the world and the people living in it. I just want to make sure the right priorities are selected.

Many of the world's scientists have named the rise in carbon dioxide in the atmosphere as the primary cause of this slight warming of Earth. They show graphs of the increase of carbon dioxide being generated and added to our atmosphere. Their information is that the carbon dioxide in the atmosphere has gone from approximately 280 to 380 parts per million, an increase of 36 percent. Most of it has shown this increase to have come over the last fifty years. Other writers have stated that the increase has been more like 38 percent over the last one hundred years. Most of it came over the last fifty years. In all cases, they blame the increase on the industrial revolution, which resulted in an increase in machinery that used some form of fossil fuels. In particular, many say the increase in automobiles in the world and the burning of gasoline or diesel fuel in these cars is the main culprit. I have discussed my view that other sources generate atmospheric carbon dioxide. They are just as significant, if not worse, than those from automobiles. The increase in humans, along with the food supplied by the animals and plants and the total infrastructure required to support this growth, has created more carbon dioxide than the exhausts of automobiles. But I am not an automobile advocate. The combination of automobiles, people, and food supplies for people may be the culprits, if there are any culprits, if carbon dioxide is the major source of global warming, and if there is global warming.

Data has shown that an automobile, on average, puts out three times the carbon dioxide as man does in a year. This is based on the fact that automobiles put out carbon. When combined with air, it provides 19 pounds (8.6 kilograms) of carbon dioxide per gallon of gas burned. When this is extrapolated across the number of cars in the world and multiplied times the number of days in the year, it arrives at a level that is three times the output that the entire world's population expels, that is, those who expel carbon dioxide rather than carbon. This is based on each automobile consuming one gallon of gas a day. However, recall the data I derived about the total calories required to support man's food needs. This required ten times the amount of man's input per day. This is mainly due to the large amount of calories required by the animals in the food loop. These same animals breathe out carbon dioxide at a rate greater than man. Therefore, if you take the carbon dioxide that man puts out in a year and multiply it by ten, you will come close to the amount of carbon dioxide being generated to support man. This turns out to be 3.3 times the amount claimed for automobiles. The automobile is not the culprit in generating enough carbon dioxide to create global warming; it happens to be one of the several and is not the biggest by far.

It is estimated that a car's exhaust of carbon dioxide results in 19 pounds of carbon dioxide for one gallon of gas. Assuming that all the drivers in the world with their 650 million cars use one gallon a day this would result in 1.235×1010 pounds per day or 4.5×10^{12} pounds per year. It is estimated that a normally active man expels the equivalent of 1.0 pound of carbon dioxide per day. There are 6.7×10^9 people on earth at this time. This would result in their total exhaust being 6.7 billion pounds of carbon dioxide per day or 2.5×10^{12} pounds per year. So this would indicate that they are about equal with automobile exhausts. However, in order to feed man the cattle, swine, chickens and other foods require ten times the amount of expelled carbon dioxide as man. Or 2.5×10^{13} pounds of carbon per year. This means that man and the food chain to supply his needs provides about ten times the carbon dioxide that car exhaust creates. So, if carbon dioxide is increasing in the atmosphere, it is more likely being caused by the fantastic increase in the population.

Keep in mind that I have not witnessed any increase in global warming. I agree that Earth has benefited by global warmth, which has been around for several thousand years. Has Earth gained an increase in temperature in the last hundred years? It is not enough to be of major concern. It is not enough to call it global warming. It might be enough to make people aware of the possibility and take preventive measures to reduce the possibility. I am in favor of reducing the use of the fossil fuels as much as possible. This is politically correct, economically correct, and environmentally correct. It is just plain good sense, but it is not an action required to prevent global warming, if this warming exists.

The problem is one of politics and economies. The people of the various countries, along with their leaders, gain economic power from their oil production, no matter what they are producing. However, there can be no production without the use of fossil fuels. We use hydroelectric power to produce electricity; and this adds little, if any, greenhouse gases to the atmosphere. But not everyone has the capability to have hydroelectric power. Meanwhile, there has been no major development of an energy source to take the place of fossil fuels. There has been a lot of talk about other powers, like solar and wind, but there has not been a real push.

Other Possible Reasons for the Increase in Carbon Dioxide

On the other hand, looking at what may have occurred over the last fifty years and where there has been a rather large increase in the carbon dioxide in the atmosphere; I find what may be a few major culprits that caused this recent increase. If we look at the time from about 1970 until the present, a couple of major events resulted in an increase in the amount of carbon dioxide created. One was caused by the worldwide actions taken to eliminate lead, carbon monoxide, nitrous oxide and others in gasoline. Catalytic converters were added to every car, and cars had to go through smog checks. This did a terrific job of reducing lead, nitrous oxides, carbon monoxide, hydrocarbons, and other gases that were dangerous to the health of humans and some animals. The catalytic converter significantly reduced these exhaust gases. It also added oxygen to the exhaust stream, turning carbon monoxide to carbon dioxide in the exhaust. In addition, other chemical reactions, along with the oxygen addition to the exhaust, resulted in some unwanted, but less dangerous gases. The recovery of some of the exhaust and routing it back through the combustion cycle to have it burned again has resulted in more efficient use of fuel and helped to reduce the hydrocarbons in the exhaust trail.

Before catalytic converters, cars had a trail of visible smoke. Now, this smoke has disappeared. Smog has been reduced in the United States. These catalytic converters and the change to unleaded gas had its good and bad points. A bad result was a dramatic increase in the amount of carbon dioxide being exhausted. The engineers who promoted this approach knew more carbon dioxide would be exhausted from the cars, but they weren't really concerned because this gas was friendly compared to nitrous oxide, carbon monoxide, hydrocarbons, and so forth. Here was an invisible gas that no one worried about. Clean air measurements in California demonstrated that the gases have improved the air quality by a dramatic amount. However, these reports on the improvement of air quality do not count carbon dioxide as a dangerous gas because it is not a listed as a dangerous gas. In fact, carbon dioxide is not considered a pollutant. There was never a drive to reduce carbon dioxide in the environment when the world was taking steps to reduce the lead content in gasoline as well as reducing carbon monoxide, and nitrous oxides coming out of the tailpipe of a car and other air pollutants. Carbon dioxide is still considered a reasonable gas if it doesn't result in global warming. Does it?

However, it is a new day. Now our focus has changed. Some of the scientists of the world consider carbon dioxide to be the culprit in global warming. One alarming figure relates to the increase in carbon dioxide since 1970. In 1968, there were around 70 million cars in the world. Today, that number is close to 650 million. This is an increase of almost a factor of ten. All have catalytic converters on them, and they are all pouring out carbon dioxide while reducing the other problem gases. So, this worldwide insertion of the catalytic converter, along with the huge increase in the number of cars on Earth, has resulted in a jump in the amount of carbon dioxide injected into the atmosphere. Perhaps the same engineers who developed the present methods that have shown a great improvement in air quality can now take a look at what it takes to reduce carbon dioxide exhaust being pumped into the air. Even though this should have shown a dramatic increase in the carbon dioxide in the atmosphere, it hasn't. This means the scientists of the world are missing something in their analysis.

Another major event that could be responsible for the increase in atmospheric carbon dioxide was the eruption of Mount Saint Helens on March 20, 1980. This eruption followed several earthquakes in the area. The initial eruption was mainly steam that added to the atmospheric water vapor. A couple of earthquakes followed. On May 18, the final blow came from another earthquake, which resulted in the biggest known debris avalanche in recorded history. This volcanic explosion eliminated vegetation and buildings over a huge area of approximately 240 square miles. For more than nine hours, a large plume of volcanic ash was ejected. It reached places as far away as Idaho, 300 miles away. The ash in Idaho was so dense that one could not take pictures. This ash fell for more than a week. The amount of dust put into the atmosphere must have created unfamiliar territory for our atmosphere and Sun to break through.

Ash in the atmosphere causes an accumulation of water droplets, which eventually had to result in abnormal weather. Cloud formation from Mount Saint Helens must have been extraordinary. I have not read any details on the amount of carbon dioxide put into the atmosphere, but it must have been a tremendous amount. However, carbon dioxide was not the big worry then. The bigger worry was if this large plume that eventually made its way around the world, although at a much reduced amount, would cause any global weather problems. The energy released was equivalent to 850 megatons of TNT, or 27,000 atomic blasts, like the ones dropped on Japan during World War II. (37) Since that time, the volcano has been active, but it has been not like the May 18 eruption. From November 1990 to February 1991, there were large eruptions of ash. On March 8, 2005, a 36,000-foot-high ejection of steam and ash could be seen in Seattle.

Wars

A couple wars did happen in the meantime. The first invasion of Iraq did not last long, but the length of time it did occur resulted in numerous bombings by our aircraft. A considerable amount of firearm and tank weaponry created considerable extra energy to be exerted. The normal things of war polluted the air. Due to the desert, there was probably considerable dust in the air to attract water vapor. And water vapor is expounded to be a worse greenhouse gas than carbon dioxide.

Next, American troops bombed and invaded Afghanistan as a result of the September 11 terrorist raid on New York City and the Pentagon. The 911 terrorist attach on the two major buildings in New York created a fantastic amount of dust of all types. The fantastic bombing of key areas in Iraq followed. We then invaded that country. This dramatic bombing and invasion released considerable energy and created a pollution of the air over and around Iraq. Various groups within Iraq attacked the oil supplies, resulting in fires and pollution of the atmosphere. These oil fires were of significance and probably added as much carbon dioxide to the atmosphere as all the other encounters in that region. Much of this internal war resulted in a tremendous amount of dust and pollution of the environment, which could add considerable water vapor in the air. I believe it added to the events occurring about that time to increase air pollution and the increase of carbon dioxide.

I did not count the war in Vietnam because it was just before the 1970 period I have used as a reference, but a good bit of any gain in carbon dioxide may be due to this war. Considerable defoliants did strip the jungles of Vietnam to provide better targets. This huge amount of greenery lost resulted in a huge loss of plants that removed carbon dioxide from the atmosphere. This loss probably was in full effect in the early 1970s. It probably should be noted as a possible addition to the other one-time events I have mentioned.

Rather than the automobile and other human endeavors, it is possible that the large increase in carbon dioxide from 1970 to the present could be partially explained by the following: the shift to the catalytic converter, the eruption of Mount Saint Helens, the first war in Iraq, the war in Afghanistan, and the second war with Iraq. These events, if not primary, added to the situation. Maybe these are one-time happenings. Maybe Earth will return to carbon dioxide in the atmosphere that is in the range of 300 parts per million. Perhaps carbon dioxide is like a time bomb. It might be that it takes time for its presence to be felt relative to an increase in the global temperature. Every event has a time factor as to how soon the event will create some related action/reaction occurrence.

Scientists have stated that carbon dioxide in the atmosphere will stay for at least 100 to 200 years. Maybe they are right. Maybe all we can expect is that any accumulative amount will have no increased effect related to the absorption of Earth's reradiation due to infrared wavelengths. Therefore, there will be no major effect on climate. If scientists are correct about the length of time that carbon dioxide remains in the atmosphere, then I would have expected there to be more than 380 parts per million. These numbers do not add up. Of course they are incorrect. The amount of carbon dioxide that man, animals, and cars have injected into the atmosphere have increased by a tremendous amount since the year 1850 and cannot be accounted for by this rather small increased amount of 100 parts per million in the atmosphere. What got rid of all that carbon dioxide over these past 150 years? There must be some other event that is taking place to neutralize the amount of carbon dioxide being generated It may be that the Earth's precipitation has washed down a large amount into the ocean carbon sink. Or there must be some other act of nature that we haven't identified. There must be a balance between what's in the Earth's oceans and the atmosphere .to maintain carbon dioxide equilibrium.

We have seen increases in the population by a factor of five since 1850, that is, from 1.25 to 6.5 billion. The automobile went from nothing to 650 million. There have been other worldwide happenings like Mount Saint Helens and several wars. Who or what can put the most carbon dioxide in the atmosphere? I am actually surprised that it has not increased by more than 20 to 30 percent. When you consider that the trees and other consumers of carbon dioxide in the oceans have essentially stayed constant or decreased while these generators of carbon dioxide have been hard at work, it makes you wonder where the carbon dioxide has gone. Scientists say the carbon dioxide in the atmosphere takes several hundred years to dissipate. Where has this huge increase gone? It brings questions as to whether we know what is happening. We know that rain reacts with carbon dioxide. Much of it is then transferred to the oceans of the world. Perhaps the rain had a greater capacity for neutralizing the effect of airborne carbon dioxide than we gave it credit for. Maybe the rain has nicely handled all the increase of carbon dioxide added to the atmosphere over the past fifty years and deposited in the oceans of Earth...

For all the carbon dioxide generated over the past fifty years, there is little evidence of it being in our atmosphere. There has been an increase, but it is little compared to the increase in people and machinery that generates it. We are missing something. Is Mother Nature playing with us again? It has been estimated, and maybe calculated that the plants on Earth take in 6×10^{14} pounds (2.73×10^{14} kilograms) of carbon dioxide per year.[73] Maybe this has been enough to keep things balanced. Maybe the increase since 1970 is when Earth's carbon dioxide generators exceeded this amount of carbon dioxide consumption. Then we saw the increase. Maybe that's Mother Nature's way of providing a correction factor. Maybe the rather small amount of carbon dioxide in the atmosphere is the answer to why we have not seen any large increase in the temperature? We are talking about carbon dioxide going from 0.0280 percent to 0.0380 percent of Earth's atmosphere. This is the increase from 280 to 380 parts per million, and some scientists are claiming that it might make a huge difference. The percent change has been significant, but the amount has not. Maybe that's the answer. Based on these numbers, we definitely have not injected much carbon dioxide into the atmosphere. We definitely have not seen a significant increase in the temperature.

---

[73] Smithsonian Release

Water Vapor

Maybe we should be talking about the total percent of water vapor plus carbon dioxide that is in the air and if that total has changed. Water vapor causes 40 to 70 percent of the greenhouse effect on Earth. Carbon dioxide causes 10 to 26 percent. Because water vapor is the greater absorber of infrared wavelengths than carbon dioxide and there is more of it in the atmosphere (5,000 parts per million to 10,000 parts per million), why isn't this discussed?

Let's assume that, in 1850, when the industrial revolution began and carbon dioxide was 250 to 280 parts per million in the atmosphere, water vapor was 1,000 parts per million. Now, when the carbon dioxide is 380 parts per million, the water vapor is down to 500 parts per million. In this hypothetical example, the total of these two greenhouse gases would have gone from 1,250 parts per million to the present value of 880 parts per million. This would result in a decrease in the infrared absorbing gases in the atmosphere, ignoring methane and the other small contributors.

Let's take another hypothetical situation of the same sort. In 1850, the water vapor in the air was 5,000 parts per million and the carbon dioxide was 280 parts per million. Now let's assume the water vapor in the air remains the same and is now 5,000 parts per million and the carbon dioxide is 380 parts per million. There would no impact from the increase in the carbon dioxide by 100 parts per million. The carbon dioxide is insignificant in both cases.

These are hypothetical examples, but it could explain why we have shown only a minimal increase in temperature. I have not seen any good comparisons of this total of infrared absorbing gases over the time period being mentioned. I have looked through the data and found no mention of the combination of the two constituents in the atmosphere. This may be due to the fact that the amount of water vapor in the atmosphere is changing all the time, making it very difficult to obtain real-time data. Reports say that water vapor varies from a level of approximately 500 parts per million to a little over 1 percent and could reach 5 percent on certain days. This is a rather large change. It is much more than the carbon dioxide varies and much more in absolute value. This somewhat reminds me of the light bulb not being able to last long because the tungsten filament burned up in the bulb. Then they added argon to the contained light bulb. It then lasted at least a year before the filament was consumed. Maybe a compensating element in Earth's atmosphere is doing something like this to compensate for the increase in carbon dioxide. Perhaps it is related to a combination of elements that impact the atmosphere, such as pressure, the amount of carbon dioxide or water, and the amount of sunlight passing through the atmosphere, the amount of photosynthesis occurring, vertical winds caused by evaporation of the ocean's waters, the atmosphere above the troposphere, the stratosphere, or other ingredients of nature. Maybe it is as simple as the precipitation that has been constant over many years and the precipitation is water vapor in its ultimate form. Precipitation can take carbon dioxide down to the waters of the oceans where it is a perfect sink. We have not seen any increase in acid rain, have we?

The Concentration on Carbon Dioxide

Without any evidence to the contrary, I believe that scientists have concentrated on the increase of carbon dioxide in the atmosphere and have not taken into account the water vapor in the atmosphere because they do not have models to show the water vapor percentage in the atmosphere at any given time. This is not an easy task to perform, even on a computer. Carbon dioxide in the atmosphere is easy to monitor because it stays constant over a period of time or changes in one direction slowly. It is easy to show that it has been increasing because it is a stable function. On the other hand, water vapor changes continually—daily, weekly, monthly, or yearly. And the change is not always in the same direction. It can go up or down at any given time because it is a balancing parameter. It balances as a function of many variables. If the oceans are heated above a certain level for a few days, then the atmosphere responds through increases in evaporation, atmospheric water vapor, clouds, precipitation, and mixtures due to winds. These are real-time events, meaning they can take this form at any time and perhaps balance things. This is one of the reasons why there is a change in where precipitation takes place in the world at any given time. Various elements within Earth's weather system move the water that gets in the clouds. If the exact same amount of water vapor were to be released each day on Earth, do you think we could tell where the precipitation is going to occur on Earth? This is a very dynamic system. Just keeping one of the variables constant, like water vapor, would result in different amounts and different locations of precipitation in the world every day. When you think of all the clouds around Earth and one may be empty of water vapor, one may have 50 percent, and one may have another amount entirely, it makes determining their daily effect at the various locations almost impossible. Mother Nature does us a good turn again.

Remember when I discussed Earth/Sun energy balance, earth's energy budget and the atmosphere, global heat balance, the ocean mass water balance, and nature's big circle? These all explain the major balances within Earth's structure. As much as I could explain many of these phenomena for which there have been many papers written, I could never explain or find enough clear data as to how the atmosphere, or water vapor, changes on a daily basis to accommodate many of these phenomena. The atmosphere responds to increases in evaporation of water, but it is not immediate. As Earth spins at 1,000 miles an hour, causing a vector that is from west to east and the heat transfer of Earth from the equator toward the poles results in vectors to the north and to the south away from the equator, these vectors combine to cause winds in directions that are the combination of these vectors. The poles also have an impact on the weather due to the jet stream. All play a part in the weather and the delay of the water evaporated. In addition, cloud formations caused by increased evaporation impact reflections of the Sun back into space. It is estimated that a car's exhaust of carbon dioxide results in 19 pounds of carbon dioxide for one gallon of gas. Assuming that all the drivers in the world with their 650 million cars use one gallon a day this would result in 1.235×10^{10} pounds per day or 4.5×10^{12} pounds per year. It is estimated that a normally active man expels the equivalent of 1.0 pound of carbon dioxide per day. There are 6.7 x 10^9 people on earth at this time. This would result in their total exhaust being 6.7 billion pounds of carbon dioxide per day or 2.5 x 10^{12} pounds per year. So this would indicate that they are about equal with automobile exhausts. However, in order to feed man the cattle, swine, chickens and other foods require ten times the amount of expelled carbon dioxide as man. Or 2.5 x 1013 pounds of carbon per year. This means that man and the food chain to supply his needs provides about ten times the carbon dioxide that car exhaust creates. So, if carbon dioxide is increasing in the atmosphere, it is more likely being caused by the fantastic increase in the population. But an increase of only 100 parts per million doesn't appear to be enough of an increase to cause a problem.

Various elements play a daily part of Earth's weather for that day. If things were not tough enough to predict, large forest and field fires during the summer fill the air with dust that ends up in the clouds and changes the way that the clouds react. The dust particles collect water vapor and form very small drops of rain that do not fall in the form of precipitation at that given time. The drops are so small that the vertical updraft caused by water being evaporated from the oceans causes an upward pressure that keeps these small water drops in suspended air. Simply, it is called a cloud. Sometimes, there is an updraft caused by direct conduction between the warm water and atmosphere above it. Perhaps this updraft goes all the way up to layers beyond our tropopause layer (about 30,000 to 50,000 feet) and works a balance. All of these functions are very difficult to model.

Monitoring the water vapor in the atmosphere is tough enough. Next comes precipitation's part, which may be even harder to measure and model. Remember the big circle, where I described the water being evaporated and taken into the atmosphere, which eventually results in some form of precipitation to return the water to the oceans? Nothing is discussed about the precipitation and its impact on global temperature. It must also be a constant. The temperature of the atmosphere above the rain clouds must receive constant pulses of heat as the clouds have their water condensed and pass the heat to the atmosphere above it. Because this energy must reflect the energy relieved from the oceans, which is a constant, it is too hard to determine on a weekly or monthly basis the effect on these huge atmospheric systems above the rain clouds.

If the precipitation is in the form of snow, then there is reflection off this snow, which impacts the amount of heat that Earth receives from sunlight during that moment in time. It is different than rain. What if it rains? Where does it rain? One element just does not determine rain. It can be delayed, moved, cooled, or warmed. Any other elements can change its direct response. It definitely will rain. When and where? We know it cannot rain more than what is evaporated. It can evaporate more than it precipitates over a short period. Then the opposite happens. We are back to zero difference between evaporation and precipitation. The two go hand in hand. The difference is the time lag between the evaporation and precipitation. As far as anyone can tell, they remain equal over long periods as was discussed earlier in this book.

Atmosphere's Size and Capability for Holding Carbon Dioxide

When you consider the size of Earth, the atmosphere is rather small in thickness and volume. Here, we have a ball that has a radius of about 4,000 miles. The radius of the atmosphere is approximately 4,010 miles if one decides to define the atmosphere as ten miles above Earth's surface. It is hardly a blip on the view from outer space. However, as small as it is relative to the size of Earth, it plays a huge part in what happens on Earth. But the atmosphere is not a constant. It expands or contracts due to what acts upon it, especially in the troposphere. If we would pick some part of the upper atmosphere that we would call out as our standard reference and then declare that this level determines where the outer part of the atmosphere is on a daily or weekly basis, then we could monitor that and see what response it has on the many other variables that comprise the atmosphere. We could measure if there was a change in this upper level for changes in the troposphere or tropopause.

But why would we want to do this? We need to find out why a large increase in carbon dioxide in the atmosphere has not really shown much of a change in temperature, if any. Maybe some part of the atmosphere expands or contracts to take in the changes in various things, such as the change in carbon dioxide or water vapor.

We could make an arbitrary decision to make 10 miles (about 16 kilometers) up as the reference level to always use when making measurements relative to the total atmosphere. By mass, 90 percent of the atmosphere is below this altitude. The atmosphere thins out dramatically above this height. Presently, the atmosphere could change in depth from 10 to 9.8 miles without anyone noticing the change as things stand at present. If we use the 10-mile point as a reference, it might help us to determine any effects that occur at that point when we have major weather changes. It might give us a better picture of any tendency for potential climate changes. However, this small change might be the response of the atmosphere to changes that occurred within the outer confines of Earth itself.

In other words, the atmosphere is comprised of several layers with huge variations in temperature within each, and each serves a function. The part we inhabit, the troposphere, is rather comfortable, having an average temperature of 59 degrees Fahrenheit (15 degrees Celsius). In the next layer, the tropopause, the temperature is much colder, typically -58 to -103 degrees Fahrenheit (-50 to -75 degrees Celsius). It is about 8 miles (about 13 kilometers) up. The next layer of the atmosphere, the stratosphere, returns to fairly nice temperatures that hover around 20 degrees Fahrenheit (about -7 degrees Celsius), just a little colder than the temperatures we live in. The stratosphere separates Earth's atmosphere and outer space.

One layer within the stratosphere is called the ozone layer. It gets its name because it is the layer of the atmosphere where the ozone intercepts the Sun's ultraviolet rays. The temperature within that portion of the stratosphere is mild due to the capture of the ultraviolet energy at this level. Remember, this layer was formed just after the Earth received its third atmosphere which contained oxygen. The suns rays split the oxygen in this upper part of our stratosphere and created ozone. This was fortunate since the ozone captures most of the ultraviolet rays that would have made live on earth impossible. This layer of ozone is about 30 miles (nearly 49 kilometers) up from the surface of Earth. It warms the lower part of the stratosphere. The top of the stratosphere is 60 miles (about 96 kilometers) up. You would not expect to find this mild temperature as you are moving outward toward space, but, as you move up through the stratosphere toward outer space, the temperature drops rather dramatically. Soon, you are in the portion of outer space where the astronauts make their temporary home. It is very cold. These layers within the atmosphere have inconsistent thickness. They vary somewhat. Personally, I believe they also respond to changes that are occurring on Earth and change to counter any major shift in climate.

I believe changes in the troposphere impact the tropopause. When I have flown in commercial jets at 35,000 to 40,000 feet, the pilot did change course to miss weather problems. You can sometimes feel the plane drop down and back up during the flight. If the weather of the troposphere impacts the tropopause in this manner, then I believe there is an affect on the stratosphere also, but it is hard to measure. The air is almost nonexistent in these upper layers, so it is difficult to measure changes in pressure. We need to be able to monitor and measure changes at 10 miles (about 16 kilometers) up with very sensitive equipment. This part of our atmosphere, the thickest layer, contains the most volume with the least density. Perturbations are hard to measure and monitor on a continuous basis when the density is light.

What is the volume of the atmosphere if we only consider the atmosphere to be 10 miles from ground level, that is, about the beginning of the stratosphere? The volume of a sphere is equal to 4/3 x 3.14 (pi) x the radius cubed, or 4.133 r^3. By multiplying 4,000 miles (approximate radius of Earth) by 5,280 feet per mile, we will get the radius of Earth in feet, or 2.112 x 10^7 feet. The volume is 3.944 x 10^{22} cubic feet. To get the radius of the 10-mile point, multiply 4,010 miles by 5,280 feet per mile, or 2.117 x 10^7 feet. The volume is 3.974 x 10^{22} cubic feet. The difference between the two is the volume of the atmosphere using 10 miles as the outer point of the atmosphere. This value is 3.0 x 10^{20} cubic feet. The amount of carbon dioxide pumped into the atmosphere per year is 1.44 x 10^{14} cubic feet. Therefore, the amount dumped into the atmosphere per year is 1.44 x 10^{14} divided by 3.0 x 10^{20} cubic feet, or 0.48 x 10^6 parts per year. This is 0.48 parts per million in a year. The present carbon concentration is 380 parts per million. In ten years, the parts per million added to this number is 4.8 parts per million. In thirty years, it would add 14.4 parts per million. These are insignificant amounts when 10 miles are used as the atmosphere limit. If I would have used half that distance, then there would be twice this concentration of carbon dioxide in the atmosphere, which would still be insignificant. I used thirty years as the upper level because the data I will be covering shows that we will be lucky in the year 2037 to have the quantity of daily use of oil available that we presently enjoy.

These layers of the atmosphere act as buffer zones between Earth and outer space. They intercept the sunrays. The ultraviolet is almost eliminated at 30 miles (about 49 kilometers). The other rays pass through most of the layers, except for clouds that reflect visible and infrared sunlight, without being impeded. Very small water drops make up these clouds. Not only do they reflect some sunlight, they also absorb infrared heat waves. Because the water vapor is variable, along with the other weather actions on Earth, the incoming absorption of the Sun's infrared portion of light also varies as a function of the amount of water vapor and carbon dioxide present in the clouds or general atmosphere. After this water vapor and carbon dioxide is heated, it radiates in all directions. We can assume that approximately half is radiated out to space and the other half is radiated to Earth. This is the same action that is described in the descriptions of global warming, where Earth radiates infrared energy and water vapor and carbon dioxide absorb it. Any increases in absorption of infrared wavelengths in Earth's atmosphere goes both ways; the infrared radiation from the Sun would be cut in half upon entering the atmosphere; while at the same time infrared radiation or reradiation from Earth is doubled. This would make the arguments for any increase in global warming due to an increase in carbon dioxide or water vapor as moot, at least for the infrared wavelengths of the sunrays.

What if the water vapor and carbon dioxide are at different altitudes? What if the water vapor is at 15,000 feet and the carbon dioxide is at 100 feet? Will this impact the incoming Sunlight? Certainly, the water vapor and carbon dioxide would impact the infrared wavelengths and the near infrared wavelengths of the Sunlight. It would absorb those wavelengths and send half to Earth and half to space. So, the infrared incoming wavelengths result in that part of Sun's energy being cut in half. Because Earth radiates only infrared wavelengths, the amount that the water vapor and carbon dioxide absorb results in half of it being radiated back to Earth and the other half being radiated out to space. This type of exchange has been occurring for at least several hundred million years, and it has been balanced. Will an increase in the carbon dioxide to today's levels impact the incoming Sun's energy enough to keep Earth's radiation and reradiation from having a major effect? So far, there has not been a major increase in Earth's temperature while all this increase of carbon dioxide has been occurring. Is something balancing the act? Of the scientists I know of, none has answered this question.

I have discussed that weather on Earth is a variable, but climate is not. Climate, under the conditions of Earth that have prevailed for the last 10,000 years, has remained stable. Even with the small ice age about 500 years ago, it was tolerable by man, and it did not cause major damage to our life on Earth. It was a slight variable. Then we returned to the conditions that prevailed just before it began. So, this can be represented as a change in weather, rather than a change in climate. This happened to be a long change in weather, so I may be stretching the facts. But no one can say this weather, or climate, change was due to a drop in the carbon dioxide level or anything like that. Maybe it was due to a drop in water vapor.

Climate Change or Recovery

Due to the variations, such as responses and corrections, that continue to happen, some functions within Earth's climate cannot be modeled accurately. These variations are part of the phenomenon that dictates that energy cannot be created or destroyed in a closed system. I have drawn a picture for you as to what has happened since Earth's birth. I selected a time when I felt that Earth reached equilibrium, specifically around the Cambrian Explosion to the present. More recently, during that time period, the evolution of man became a reality. Man has had an impact on Earth. Earth has responded through Mother Nature and provided a balance that we presently enjoy. There has not been a major change in temperature or climate over the past several hundred years. I do not consider a rise of a degree over a long period as a global warming issue. It is a slow recovery from a mini ice age that occurred about three hundred years ago. If we took the temperature of this mini ice age out of the equation then the change in temperature is only about a half of a degree Fahrenheit (little over a half degree Celsius). This is rather insignificant.

One of the explanations for the lack of climate change relates to where and when the Sun heats Earth versus where and when Earth radiates its heat back. Because of Earth's shape, the Sun radiates a major percentage of its solar energy to Earth around the equator. The equator is that much closer to the Sun than the poles. This radiation is "line of sight" transfer of heat Whereas, Earth's reradiation is not line of sight; it radiates from the complete sphere 24 hours a day; from a much cooler planet and from all over the planet due to Earth's radiation coming from infrared that is mainly radiated by Earth's oceans and, more recently, from the large number of people on Earth. I am not saying that radiation from Earth is evenly proportioned from our globe, but I am saying that it is not as focused as the Sun's energy. Through heat fluxes and winds, the Sun's energy is spread out continuously with its focus on the equator and the gradient of temperature going from the equator toward the poles. This takes time. However, because it is occurring constantly, one can think of it as occurring at the same time. In other words, each day that the sunrays hit the equator, it can be considered as spread during that same day to the other parts of the world. A snapshot of the world and the temperatures of the ocean on one day will look the same if taken a week later. This being the case, you can consider this as the constant Earth's temperature at any location, and, in general, it is. It is not the same water or air in New York City as what hits the equator and evaporates that day, but it can be considered as such. The key is that approximately 1,366 watts per square meter hit the outer level of our atmosphere. Of this amount, 19 percent is absorbed in the atmosphere, 30 percent is reflected by clouds, and 51 percent makes it to earth. Of this 51 percent, some is reflected back to the atmosphere and space. The Sun sends its rays on a line of sight down on a section of Earth for a split second. This is one quarter of Earth during that split second since line of sight of a sphere only shows one quarter of a sphere. Since earth is spinning we can approximate that quarter of the Earth receiving this sun for six hours of the day; since it will receive this full treatment for only six hours of the twenty four hour day during one revolution of earth. So one fourth of Earth is exposed fully for about six hours at a time and then goes into gradual shadow of the suns ray and finally that portion of earth is in shadow for approximately six hours and then partial exposure for six hours. The energy is dissipated via Earth's spin, winds, heat fluxes that move the energy around, evaporation, diffusion, conduction, and ocean currents. When that part of Earth is exposed to the Sun on the next day, it has averaged approximately 340 watts per square meter of Sun's heat for that day. A given portion of Earth has received an average of 340 watts per square meter in one day. Meanwhile, the reradiation for that section of Earth is on a twenty-four-hour basis. In fact, it radiates more during the night than during the day. This occurs all over the planet. We know the Sun does not only heat that one-fourth of Earth. The other three-fourths reacts the same. The radiation of the infrared wavelengths from the oceans is probably the main radiator from Earth. It does it twenty-four hours a day as each part of Earth is heated by the Sun as it spins. The Earth radiates the infrared wavelengths while the Sun is striking it and during the night. Maybe the atmosphere's absorption of the infrared wavelengths is not as high during the night as it is during the day and we will not have as much infrared being absorbed by the water vapor and carbon dioxide. I have not seen a paper on the absorption of infrared wavelengths versus temperature for water vapor and carbon dioxide over a twenty-four-hour basis. This might explain the reason why we have measured more (on a percentage basis) carbon dioxide in the atmosphere, but it hasn't shown the temperature increase (global warming) that scientists have been, and are, predicting. Keep in mind that the Earth is radiating much of its energy into cold space during the eighteen hours each portion is not bathing in sun. The difference in Earths temperature of ~278 Kelvin and the void of space it is radiating into during most of the day is very dramatic in difference... This void of space is approximately 4 degrees Kelvin. So the difference between Earth's temperature and the void of space is more than the 20 times that the sun is hotter than Earth. The ratio between the temperature of Earth and this void is approximately 70 times. One would expect a great amount of heat is emitted to this cold space each day. I have not seen the figures on this radiation to cold space.

When you consider the large increase of humans, animals, automobiles, and industrial production, the increase in the amount of atmospheric carbon dioxide has not been that great. Scientists haven't been able to explain this fact. If carbon dioxide stays in the air for a hundred or more years than I believe the increase we have seen over the last one hundred and fifty years has not been a true indicator of any increase in global temperature. Something must be offsetting the relatively low change we have seen in carbon dioxide during this period of time. We also have not seen an increase in temperature that reflects this increase in carbon dioxide. Keep in mind the temperature increase being touted began at the end of the Mini-Ice Age we saw in AD1500 to the early 1800's. So, it is recovering from a low temperature to one that is more normal.

Perhaps the tendency for an increase in atmospheric carbon dioxide causes an increase in plant growth or the plant's absorption rate of carbon dioxide? With the increase in humans and animals over the last hundred and fifty years, there has been a greater demand on required oxygen. Have we seen any decrease in the amount of atmospheric oxygen? No trend shows this. What if the greater amount of carbon dioxide has resulted in a faster new growth of greenery or, essentially, the increase of oxygen generated? New plant growth consumes carbon dioxide faster than old growth. Maybe nature's way of balancing the fantastic increase in people and animals over the past one hundred years, along with a greater generation of carbon dioxide, has been met by this faster growth of greenery. Keep in mind that man has created vast agriculture growth. There are more and more fields of corn, wheat, rice, rye, and oats over large expanses of space, especially in the Midwest states of this country. These large investments in the growth of plants must have had an increase on the amount of carbon dioxide consumed. Perhaps we are eating ourselves out of the carbon dioxide issue? What if the generation of added oxygen has made the increase in carbon dioxide a moot point? Remember when I discussed the history of the planet and reviewed the heating of Earth when the atmosphere was primarily carbon dioxide, nitrogen, and a few others of the heavier gases? Oxygen was not included. At the time, there were no oxygen respiration types of life on Earth. Then photosynthesis stepped in. Plants grew and could eventually convert Earth's atmosphere to an oxygen/nitrogen atmosphere. In fact, I can assume that, when the amount of carbon dioxide reached very low levels in the atmosphere, evolution required that animals drag themselves out of the oceans and start breathing out carbon dioxide and breathing in oxygen. If the carbon dioxide in the air drops below 50 parts per million, plants die. I would assume that evolution took its rightful place. The evolution of oxygen-breathing animals and carbon dioxide generators had to evolve. If not, plant life on Earth would have been eliminated. Without plants, you can forget about animals and human beings.

When the Cambrian Explosion took place, there was about 2 to 3 percent of the present level of oxygen in the atmosphere. Animals requiring oxygen began to appear on Earth. I am going to hypothesize that the major exchange of carbon dioxide for oxygen by the plants and animals at that time was done at a distance. I mean that the plants were here on Earth. The few animals and humans on Earth got their oxygen indirectly from the distant atmosphere. Let's say 5,000 to 10,000 feet in the atmosphere. Now let's hypothesize that, as Earth gained a much greater oxygen atmosphere and as it became more populated with man and animals, the exchange happened on a more local level, specifically a much closer level. Man and animal were right next to the greenery. There were so many of them that nature did not require them to take it from the more distant atmosphere, as I chose for this example. They could take it directly from plant to animal at the very local level of the atmosphere. This type of transfer may have been necessary to satisfy the 6.5 billion people and the fantastic number of animals on Earth, including the ones that are not a part of our food chain.

With this example, man's need for oxygen is ready at hand. The plants' need for carbon dioxide is also accessible. If a scenario such as this occurred, then it might explain why the atmosphere has only shown a relatively small increase in carbon dioxide. All the carbon dioxide that we know is being generated by man, animals, fossil fuels, and the machinery of man. All of these increases should have shown a bigger increase in the level of carbon dioxide in the atmosphere. We may have a different system to analyze then previously existed.

What if the increase in carbon dioxide does not have a linear affect on the absorption of infrared wavelengths? That is, doubling the carbon dioxide in the atmosphere does not double the absorption level of the infrared and near infrared energy levels being radiated from earth. This might explain why the increase by 20 to 40 percent of the carbon dioxide in the atmosphere over the last one hundred years does not result in a significant, if any, change in Earth's temperature or climate. I have experienced many engineering functions that react rapidly in the initial time period. Then they level off as time proceeds. This might definitely be the case because the amount we are talking about is in parts per million. Maybe going from 280 to 380 parts per million of carbon dioxide in the atmosphere does not result in a temperature increase accordingly. Maybe this change only changes the absorption rate of infrared radiation from Earth by 1 percent. Maybe this type of situation exists until the carbon dioxide level becomes around 1 percent in the atmosphere or similar to the level of water vapor. Then we may see a dramatic impact. It might be that the change from 280 to 380 parts per million is sort of lost up there, where the nitrogen is 78 percent, the oxygen is 21 percent, and the water vapor is between 1 and 5 percent. At those altitudes, the weather is such that it keeps mixing the gases so there is only a few times when Earth's radiation of infrared heat energy even responds to the difference between 280 and 380 parts per million. This reminds me of the stock market. Stocks with high numbers and change two points in a day show almost no percentage increase. Stocks with low numbers and change two points show a huge increase, percentage wise. These carbon numbers are small. A small increase shows a major percentage change. But it is not the big percentage change that is important. It is the actual amount of change that could make a difference or not.

I do not believe the carbon dioxide in the atmosphere is a homogeneous blanket completely around Earth at 380 parts per million. In fact I believe that the atmosphere is so huge and affected by temperature shifts that there is probably a major difference in the carbon dioxide in the Southern Hemisphere as to the Northern Hemisphere. I do not know if the measurements we made 150 years ago or the 100-year-old samples we took from ice are representative of the same measurements we presently take of our atmosphere. In fact they may have been taken from ice in the polar region and we are measuring today's carbon dioxide in Hawaii. This would not be a good comparison. We should definitely be better at it now then we were then. Maybe we should take a sample of ice and compare it against that of the years past. Maybe the method used in Hawaii that we rely on needs to be reviewed. Maybe some correction factors should be added. Even though they are doing a great job in Hawaii of monitoring the carbon dioxide in the atmosphere, maybe they should measure the percent of water vapor at the same time and on a continuous basis. I believe the combined level of carbon dioxide and water vapor in the atmosphere determines if there is an effect. Because Hawaii has been the one monitoring the carbon dioxide and has shown the increase to 380 parts per million, they should be the ones to tell us if they have seen a change in climate in Hawaii over the past fifty to one hundred years.

I also believe the effects could be incremental. Nothing is staying put in this scenario. Change may occur one morning and be reversed by the afternoon. There may be changes one day and not the next. If this is the case, then, incrementally, there is a change in one direction for one of these points in time and a change in the other direction at another point in time. So, incrementally, things may be changing and unchanging. Like calculus, when you sum up all the incremental changes. I realize what I am discussing is weather rather than climate, but there should be a point taken that relates to the past besides carbon dioxide. What you have is climate if it is over a long period. It is a change in the weather if the time is short. What makes this an incremental issue is that the one large absorber of infrared heat energy is water vapor. It keeps changing in all directions every minute of the day. In addition to that, it makes the level of carbon dioxide look miniscule. It just may be the heavy weight, and it has been rather constant over the last fifty years.

If you review the graph that shows the wavelengths of light that come from the Sun and pass through the earth's atmosphere, a considerable amount of infrared or near infrared passes through our atmosphere. The water vapor and carbon dioxide in the atmosphere must reduce it. If it follows the absorption described for these greenhouse gases, it absorbs these wavelengths. Then it radiates this heat in all directions. Let's assume that approximately 50 percent is radiated down to Earth and the rest is radiated back into space. As there is an increase in carbon dioxide and water vapor in the atmosphere, I would expect this would result in more absorption of the Sun's infrared light with more being radiated back to space. With this scenario, not only does the greenhouse gases absorb radiation of the infrared and near infrared from Earth, it would also attenuates the incoming Sun's light by this greater percentage. So, with this scenario, less Sunlight passes through the atmosphere. It may offset the increase in radiation from Earth of these wavelengths. In addition, any increase has a tendency to build more cloud cover. This may result in additional reflection of the incoming sunrays.

Compared to Earth's percentage of infrared being reradiated, there is a large difference in the percent of Sunlight that is in the infrared. However, the Sun's percentage of infrared is a percentage of a lot more energy than what is radiated from Earth. All of the powerful sunrays hit on the outside of the stratosphere. Models and reports indicate that maybe 50 percent make it to Earth's surface. With an increase in water vapor and carbon dioxide, maybe only 45 percent reach Earth. Earth now is receiving less energy due to this attenuation. As a result, with this scenario, we have a cancellation of the increase of Earth's added infrared heat.

Incidentally, I could have made an argument that is in the opposite direction. Maybe the 5 percent reduction of the 100 percent of the sunrays that hit the stratosphere represents an insignificant decrease in the energy that passes through and hits Earth. The short wavelengths represent the higher energy. Maybe the attenuation of the total energy is only about 1 percent while the reradiation increased was due to the higher amount of water vapor and carbon dioxide in the atmosphere that absorbs Earth's radiation and the amount being reradiated back to Earth. This would have a tendency to cause the temperature to rise with all other things being equal. However, we have not experienced a significant rise in temperature.

Are the Scientists Models Wrong?

Much of the picture and model of Earth's climate provided by our scientists proceeds with the assumption that any additional heating of the atmosphere via the reradiation from earth that is caught in the carbon dioxide and is 50% radiated back to Earth results in additional heating of the oceans waters, particularly the surface heating of the oceans. Scientists claim that all this extra heating is only on the surface and evaporation will increase. This starts a feedback mechanism that supplies more water vapor to the atmosphere. This causes more precipitation, thus starting a runaway situation that eventually results in ocean levels increasing. The added surface area of this increase in ocean levels results in additional evaporation. This runaway situation continues to grow even faster. Where is all this energy coming from? We cannot create energy. The Sun's energy is a constant applied to Earth. This is where we must account for it. We must account for increases in the functions that control our climate. Remember the ocean mass water balance and the big circle I described previously. If scientists are going to claim that this runaway situation is going to exist, where do we take the energy from? The ocean mass water balance? The big circle? What changes in these constants that we have understood for many years? Does something change in these to give us the extra energy? Does something change in these to keep the energy constant? When we increase the amount of heat being reradiated back to earth due to these greenhouse gases in the atmosphere, perhaps it can only handle a certain amount or volume; if this happens perhaps the extra infrared waves of heat transfer all their extra heat through the atmosphere to outer space.

Earth receives its energy from the Sun. Energy is a constant. Keep in mind that the Sun's radiation is made up of short wavelength, high energy wavelengths. Perhaps the longer wavelengths that make up infrared heat that is reradiated back to Earth from the water vapor and carbon dioxide does not have high enough energy to make an impact on the oceans? We know this longer wavelength of heat energy is much lower in energy. Maybe evaporation from infrared wavelengths does not increase at the rate that these forecasts would suggest? Evaporation depends on many things. All relate to the use of energy derived from the daily Sun. Maybe the attempt to increase evaporation by increasing the amount of reradiation energy does not happen that way. In order for more evaporation to occur, there must be an avenue for the Sunlight or reradiated infrared wavelengths to strike the water. The Sun's energy is a radiant one. Radiant energy is "direct line of sight" type energy. It must see what it is about to heat. Reradiated infrared heat is not a radiant heat. It spreads out and heats in general. It therefore acts differently on the oceans surface than the sunlight. Relative to the Sun's short waves of energy, as evaporation occurs, it cools the water's surface left behind. In order for an increase in evaporation, the water that has just been evaporated must clear out of the way more rapidly to make room on a faster basis for the radiant heat to see the water's surface. If not, the added heat is hitting the water vapor that has just been evaporated because it may be providing a temporary shield over the water it just left. If this were the case, the Sun hitting the vapor results in an attenuation of the heat that could strike the water. There may be a time constant involved in the rate that evaporation can take place. Hurricanes present a good example of this phenomenon. If the hurricanes move fast, they can keep building energy. However, if a hurricane moves slower, it soon collapses because it must move fast enough to get out of its own way. It must move fast enough to expose new warm water because the water's evaporation by the hurricane causes the water to cool. If the hurricane does not move, it is standing over cool water that cannot recharge the hurricane's engine. So, in order for the hurricane to gain more energy for its engine, it must move to waters that the Sun is heating and has not evaporated yet. A time constant is involved for the heating and removal of the evaporated surface water before the next cycle of heating and evaporation can begin and be completed. I have not seen any data on this heating-evaporation time constant. That must be known to show it is feasible for a large increase. There may have to be an increase in the wind to clear the evaporated water vapor and expose new water to be heated and evaporated. There are more variables to consider than just heat and evaporation.

We have not experienced a global warming in our lifetimes. We do not know all of this information. Computer programs are only as good as the data entered. Without any background, it is hard to model. The models are based on a learning curve experience. The Sun only has three significant places to heat: land, air, and water. It is constant. Maybe the added heat goes to deeper depths in the oceans and away from the surface. Maybe this results in added photosynthesis taking place from the deeper parts of this penetration, resulting in an increase in carbon dioxide absorption. Maybe the time constant relative to the fluxes, evaporation, and ocean currents cannot match what the reradiation is trying to accomplish and the water surface cannot respond fast enough to have the extra evaporation. What if the amount of water that is evaporated and precipitated stays constant even when the reradiation is trying to increase? Maybe that is why the increase in carbon dioxide has not really impacted the climate.

The little change in temperature over the past century does not indicate an increase in energy that has been created. What do our models say about the precipitation during all this drive to obtain new energy? Does the precipitation cool everything down, like it has for many centuries? Does it get us back to our constants and preserve our scenario that energy cannot be created? Maybe there is a change in the size of the jugs of clouds that holds the water vapor before releasing it in the form of precipitation as a reaction to any change in the greenhouse gases.

An Earthborn Energy Separate from the Sun

To my way of thinking, the energy that the Sun supplies to Earth and the amount we send back to space has remained constant in our use of the Sun's energy. However, we found a new energy source, an earthborn source of energy that was due to the crash of a meteor into Earth 65 million years ago. This energy was stored and just recently provided the oil and its energy that it generated as a result of that crash, as well as other energy stored in the fossil fuels of Earth. It is being used up rapidly, and it may be increasing the amount of heat that escapes from Earth over these past hundred years. But the oil will be gone in another thirty or so years. This earthborn energy provides a significant amount of energy, that is, extra energy besides the energy that the Sun supplies. A review in 2006 of the amount of earthborn fuel energy used in 2004 shows that it supplied 15×10^{12} watts of power per day, or 3.6×10^9 Calories per day. This is 1.125×10^{17} Calories per year. In 2004, fossil fuels supplied 86 percent of this, or 9.7×10^{16} Calories. The remainder was hydroelectric, nuclear, and wind/solar/wood for the remaining. The following table shows the breakdown of these earthborn energies: (40)

Energy Source	Energy in Calories
Oil	4.3×10^{16}
Natural Gas	2.62×10^{16}
Coal	2.87×10^{16}
Hydroelectric	0.7165×10^{16}
Nuclear	0.7165×10^{16}
Geothermal, Wind, Solar, Wood	0.096×10^{16}
Total	**1.13×10^{17}**

Table 6. Breakdown of earthborn energies.

This total of 1.13 x 10^{17} Calories is an earthborn-supplied energy that includes 9.97 x 10^{16} (almost 1 x 10^{17}) that came from the meteor crash on Earth 65 million years ago and other forms of vegetation that lay in the ground and was converted to fossil fuels. This compares to 1.314 x 10^{21} Calories that are received from the Sun each year. Compared to the Sun's supplied energy, this is 86 x 10^{-6} or 86 parts per million. We think of it as a large amount of energy, and it is. But it is rather insignificant compared to the Sun. Notice that the non fossil fuel energy is very small compared to the fossil fuel, being about 10%. We must do more to raise this level and lower our dependence on the fossil fuels which you will see I am predicting won't be around very long. However, the use of fossil fuel is increasing dramatically due to its many uses. Of the list shown, various countries are taking steps to increase their use of nuclear power, solar power, wind power, and natural gas. These will be needed to replace the significant use of oil. The number for oil will double by 2010 over the 2004 use. This will be the peak oil use, and will start a major problem.

The Answer

If someone does not explain how the percentage of carbon dioxide has increased by a significant percentage but the temperature has not increased over the last fifty to one hundred years, then they are incorrectly assuming the relationship between the two. If the answer is that there is a long time lag from the time of increase in carbon dioxide in the atmosphere to when it causes an increase of significance in the climate of our planet and if they can prove it, fine. If the answer relates to my point that it is the total amount of greenhouse gases in the atmosphere that causes the problem, then I would like to know if the water vapor is Mother Nature's way of correcting for this issue. If the fact is that an increase in carbon dioxide causes the Sun's incoming infrared wavelengths to be attenuated by the same amount as it increases the radiation from Earth, then I would accept that. If the fact is that the radiation of Earth's infrared energy during twenty-four hours from all over the planet offsets the increase in the carbon dioxide in the atmosphere in some manner, fine. If the answer lies with the fact that there is no linear relationship between an increase in carbon dioxide and a resulting increase in the absorption of infrared radiation, that would be a satisfactory answer to the problem. If the increase in carbon dioxide also results in the increase of the growth rate of greenery or an increase in its absorption rate to offset this increase, then that might explain the issue, and we would expect it to slow down. If the increase of carbon dioxide in the atmosphere just relates to its small percentage of the atmosphere and no major increase in absorption can be expected, then that could be the explanation. If the amount of precipitation worldwide has not had a yearly average change, then there is no impact by the increase in the carbon dioxide. If there is some other explanation as to why the temperature or climate has not shown an increase while the carbon dioxide percentage increased fairly dramatically, then I am sure the world awaits the answer. If the measurements of ice samples from 150 years ago were taken in a different part of the Earth than today's measurement and they do not represent the measurements we are making in Hawaii, then that would explain the difference. If the absolute amount of 100 parts per million increase is proven to be too insignificant to cause an increase in the global temperature, then I will acquiesce. If reradiated infrared energy is so low in energy that any increase in infrared absorption will not result in an impact on the oceans temperatures, than that would be the answer. If, as I will, demonstrate there is no relationship between the increase in the percentage of carbon dioxide and climate, then I will accept that

I have described the huge percentage increase in carbon dioxide generators over the years, but we have not seen much of a numerical change in the atmosphere's amount. This indicates that something offsets the generation of carbon dioxide. I have shown that, with the amount of carbon being dumped into the atmosphere, the increase in carbon dioxide for the next thirty years will be insignificant if we consider the atmosphere as that part that is ten miles above Earth's surface. I believe 52,000 feet is a reasonable assumption for the part of the atmosphere that reacts with our planet's weather and climate systems.

The Earth's Ocean Levels

There have been predictions of Earth's oceans rising to a level that would cause many cities in the world to be underwater in twenty to fifty years. This was based on the assumption that the carbon dioxide increase would cause global warming and melt the ice on the poles and other places like Greenland. Historically, sea levels have always increased between ice ages. Sea levels have increased by more than 300 feet over the last 18,000 years. (51) This is way before man and his generation of carbon dioxide existed, at least in any number to be significant. With the increases in carbon dioxide over the past century, one would think it would have had some impact on the levels of the oceans by this date if the predictions of scientists are factual. I have read some reports that this has happened. However, there is one major report that I read within the past two months that reported no rise has been seen on the water level standard in either Denmark or Sweden. I do not have the paper, but I know it was one of these countries that monitors the level of water on an annual basis and has done so for more than a hundred years. And they have not seen this level increase.

This is an element to examine. If their data is correct, then it fits the pattern of the carbon dioxide situation that I just discussed, and Earth has not shown an increase in the ocean level. Due to the tectonic plates, land is constantly shifting. These shifts result in ocean levels seeming to increase. Maybe the ocean level is not increasing. Instead, the land is shrinking along many coasts and not on others. This has been true over time.

Monitoring the Ice

I will say one thing about monitoring Earth's change in climate. The area to monitor is the ice of the world. The melting of ice is the most sensitive indicator of either climate or weather changes. Consider that it only takes 80 small calories of heat to melt one gram of ice to ice water at 32 degrees Fahrenheit (0 degrees C). This shows how sensitive the melting of ice is to a small energy and resultant temperature change. When you realize that a man has to take in 2,000 Calories each day to exist and replace the 2000 he burns off each day and those 2,000 Calories can melt 4,400 pounds (2,000 kilograms) of ice, then you know how sensitive an indicator the melting of the ice is. For a world of people numbering 6.5 billion, then the Calories lost by them each day is 1.3×10^{13} Calories. This is enough to melt 2.86×10^{13} pounds of ice per day. This is the melting of 14.3 billion tons of ice per day, and I only counted man's Calories. If you also consider the factor of ten times this for the other animals that supply food for man, you are talking about 143 billion tons of ice per day being converted to water at 0 degrees Celsius (32 degrees Fahrenheit). This demonstrates how sensitive the melting of ice is to temperature changes, and this is without a change in temperature (water has a unique capability. It takes this many Calories to melt the ice to water, but the water remains the same temperature of the ice. To change the temperature of the melted water then takes one Calorie to raise one kilogram of water by one degree Celsius.) These two things could occur and not show a change in temperature. This does occur every day, and we have not seen any temperature changes. Of course, this melting will take place at the edge of the ice where it meets the water. I have read some articles that say the edge of Antarctica is melting, but the center of the ice is actually increasing in thickness. If this is true, then there may not be an increase in the lost of ice. It is just being displaced. This would explain why scientists in one of the Scandinavian countries have not seen an increase in the water level of the ocean.

It would be interesting to know if there has been a decrease in the temperature of Earth's oceans due to any melting ice. Theoretically, ice water and ice remain at the same temperature when they coexist. I would not expect any temperature change of the oceans that are up against the ice of the poles. In fact, it would not take much change in the other direction for this water to turn back into ice. It only has to lose one table Calorie for 2.2 pounds (one kilogram) of water at 32 degrees Fahrenheit (0 degrees Celsius) to turn back into ice. This is why it is so easy to go in and out of an ice age, but it depends on the significance of an ice age. As witnessed between 1500 and 1700, there was a mini ice age. But this type of ice age is tolerable. However, if this melting continues for any length of time, the currents of the ocean would tend to mix their water with the ice water that is not at the edge of the ice. Now the warmer water would transfer its heat to this water. There should be a resultant cooling of the oceans as they lose their heat. This loss of heat is taken up in a temperature gain in the cold water and a drop of temperature in the warm ocean water. Overall, the oceans should show this cooling effect. The result would depend on the volume of ice and the resultant increase within the volume of ocean water.

Let's take this to the extreme where all the ice melts in the world and the temperature of the oceans drop by some amount. Even though it may be a small change, the density should increase. If the density increases, it should reduce in size. If it reduces in size, it should not cause any major increase in the water level in the world. We would not see any big increase in coastal cities being overcome by the rise in the ocean's level. This depends on how much volume is included in the present ice glaciers around the world, and how much volume of water they would displace upon melting. If the oceans have increased 300 feet over the last 10,000 years there should be some evidence of this. Do we have any evidence that there was exposed land on the coasts that is now covered by this water? In other words was the coast of the Netherlands (Holland) once above sea level and touching part of England, and the same hold true for New Orleans?

Ice does have the enormous capability of absorbing heat, causing it to melt, but it does not change the temperature of the ice water it converts to the water at 32 degrees Fahrenheit (0 degrees Celsius) until additional heat energy is applied. This may be why we see evidence of ice melting, but no change in the temperature of the water or any major changes in the air temperature. One table calorie can melt one kilogram (2.2 pounds) of ice into ice water, but the water stays at 32 degrees Fahrenheit (0 degrees C). We could go along thinking that there must not be any increase in Earth's temperature because Earth's ocean's temperatures have not changed. But this might be normal since there is no change until the ice all melts. After melting, it only takes 1 table Calorie to raise 1 kilogram (2.2 pounds) of water 1 degree Celsius. The melting of the ice, but no change in water temperature, might be a local happening in countries where there is a great deal of ice and it is melting. It might show a rise in water without showing a rise in temperature because an increase in the local temperatures might have caused this change. For example, the countries around the poles might see the ice melting, but they might not see any other evidence that we are having global warming. For example, this could happen in the Scandinavian countries around the Baltic Sea.

I remember reading about Earth's differentiation, where it squeezed out much of the water that makes up our planet through volcanic eruptions or actually squeezing the water out of cracks in Earth's mantle. The report indicted that Earth is still settling. It was possible that some water might actually return to Earth's mantle. If this were to be the case, it might offset any height to be gained by the oceans due to any increase in Earth's temperature. This might be Mother Nature's way of handling this problem. Any change in the average temperature of Earth's oceans must result in some physical changes to our globe other than water rising and dropping many coastal cities below water.

Let's look at another assumed scenario. What if the increase in the ocean's density due to a drop in the average temperature as the ice melts causes it to weigh more than it has been experiencing in certain locations? This might result in the ocean floor to be decreasing in heights at various places in the world's oceans, thus allowing more water to be held without an increase in water height. This is a way out fantasy, but I am looking for one of Mother Nature's little secrets again.

What if the decrease in the ocean's temperature due to the melting of the ice results in a more stable growth of the ocean's plankton and other sea anomalies such that there is a resultant increase in the carbon dioxide receptors in the world's oceans? It would be enough of an increase to account for why there has been such a small increase in the atmosphere's carbon dioxide.

We recently had a hurricane heading toward the Hawaiian Islands. At first, they thought it was going to be a Category 4 or 5. Then they changed that forecast and said the water temperature was too cool to sustain this hurricane at those levels. Eventually, the hurricane was reduced to a Category 2 and missed Hawaii altogether. They indicated the water temperature was cooler than normal. Here is the place where the monitoring of the atmosphere's carbon dioxide level has shown it to be 380 parts per million. I would bet that they have not seen a climate or temperature change in Hawaii. I would estimate that the water vapor level around the Hawaiian Islands is so high that it controls the temperature in that area. Any rise in carbon dioxide is insignificant. When you realize that a change in carbon dioxide in the atmosphere around the Hawaiian Islands is only changing by approximately 100 parts per million over the last fifty years, it is probably insignificant in relation to the water vapor in that area.

Global Warming Summary

I obviously have not experienced, personally or through reading reports, that there has been either a noticeable increase in the planet's temperature or the rising of the oceans, let alone a global change in climate. When I read through reports on the temperature variation over the past 1,000 years, I find graphs that range from an increase of +0.6 of a degree Celsius (1.08 Fahrenheit) to a decrease in the temperatures of -0.7 degrees Celsius (-1.26 degrees Fahrenheit). I have provided one of those graphs.

If this were a graph of temperature over time on a scale that was + or − 5 degrees Celsius (+/-9.0 degrees Fahrenheit), you would hardly see this dip down and up over the thousand of years. This graph looks like the temperature moved up about 0.8 of a degree Celsius (1.44 degrees Fahrenheit) over the last 200 or 300 years, which includes the low temperatures of the little ice age.[74] Going from this below normal temperature to one that is above normal temperature makes the increase look larger than normal. If you look at the time from about 250 years ago, it has increased by about 0.5 degrees Celsius (0.9 degrees Fahrenheit).

---

[74] marshbunny.com/mbunny/sidetrin/hurricane/storms.html

According to this graph, the little ice age took place between 1500 and 1700. This was an excursion downward of about 0.6 degrees Celsius (1.08 degrees Fahrenheit).

Did anyone claim there was an ice age coming and it would destroy the people of the world? Of course, there was not much in the way of media back then to transfer this message around. In fact, they probably did not graph it. They probably only complained that it was a cold winter, especially in England. This graph shows a normal temperature change of about -0.2 degree Celsius for a long period. At this time there were about one fourth of the people on Earth then there are now. I would imagine that Sir Isaac Newton really did not enjoy the temperature when he was doing his mental thoughts on the various laws of physics. For that temperature, one can always put on a sweater and light a fire. In fact, since there were no light bulbs in those days they had to light a small oil lamp to see.

While people might not have declared an ice age coming then, we did in the late 1970s. In 1978, there were many articles about the cold temperature in different parts of the world. They were starting to say that an ice age was coming; and of course, there wasn't. It did not take long for people to start saying there was global warming coming as well. It is the nature of people to try to diagnose conditions and then make wild predictions without enough evidence. In this case it may have been worse since they not only said there would be global warming due to the exhausts of automobiles. This statement, without evaluation of all carbon dioxide from the large increase in people was like a jerking knee. Many people and countries have been studying the possibility of global warming. The countries that have signed the Kyoto Protocol are proof enough. Incidentally, I have looked into the situation because I was—and am—concerned. We should be doing something to prevent the buildup of waste of any type. As a result, this may allow us to have the fortune to have oil around a little longer. It is the only mobile fuel with energy enough in a can of gasoline to perform functions that no other mobile or portable fuel can.

This graph, from *Temperature* by John Baez, shows the temperature over the last 10,000 years.[75] It looks like it was a lot colder, and it has now settled down to a nice temperature. What we are seeing is a normal event that has nothing to do with what we humans are doing, with or without oil.

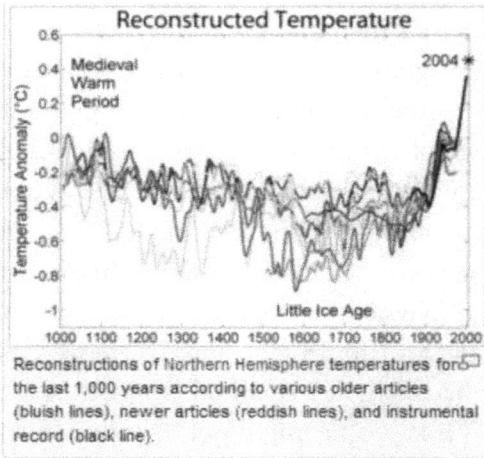

lispute   [edit]

nce of the scientific
graph centers on its use as
or anthropogenic global
3 reconstruction was
in the 2001 United Nations
nel on Climate Change (IPCC)
eport (TAR) and as a result has
d in the media.

J on technical aspects of the
a sets used in creating the

Reconstructions of Northern Hemisphere temperatures for the last 1,000 years according to various older articles (bluish lines), newer articles (reddish lines), and instrumental record (black line).

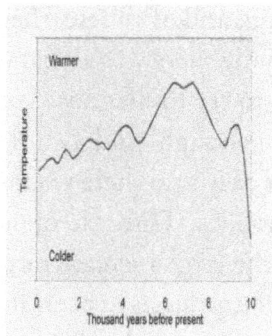

1900–Present

We should review what caused the temperature to start increasing around 1900. The industrial revolution cannot explain it. There was not enough of an increase in the use of fuels as early as 1900. Even

---

[75] World-builders.org/lessons/less/biomes/SunEnergy.html

though the industrial revolution can be considered as starting back then, there were not any major increases in what man was doing to cause this trend. However, there has been a slight increase in temperature starting at that time. World War I resulted in a large expenditure of energy from 1916 to 1918. From 1939 to 1945, World War II saw a huge expenditure of energy. The Korean War and Vietnam War again showed major expenditures of energy. There was the conversion to catalytic converters in the 1970s. The eruption of Mt. Saint Helens followed, generating considerable carbon dioxide.

On the other hand, I see no correlation between the carbon dioxide in the atmosphere and an increase in global warming. If one exists, it is of a secondary order effect. The slow temperature change since 1900 may just be the recovery from the mini ice age we had between 1500 and 1700.

Earthborn Energy

Earthborn energy was in the form of oil and other fossil fuels, providing man a major advantage. Now, not only did the Sun provide continual energy, man had the unused, earthborn energy (entropy) of the dead dinosaurs and other decomposing material that was stored over that length of time. During that period, the bodies decomposed and went from one type of entropy to another form of entropy. And did man make good use of this energy! Needless to say, as man uses the energy from something that used to be unusable, that something loses its entropy while becoming useful energy.

The key in this critical situation is that most entropy on Earth is the result of the Sun's energy not being efficiently used. Or maybe man, during the course of a normal action, does it inefficiently. As a result, there is in an increase in entropy. In this particular case I am discussing, the entropy was not from misuse of the Sun's energy. It was unusable energy tied up from the crash of a meteor upon Earth 65 million years ago. This is extra energy added to Earth from a meteor crash. Now we can put it to use. Isn't Mother Nature wonderful? She saved all this energy over that many years so we could have it available since the middle of the nineteenth century.

When Colonel Drake drilled the first commercial oil well in Titusville, Pennsylvania, in 1859, it was the beginning of the commercial oil industry. Oil had been found oozing out of the ground and was used before this, but it was insignificant compared to what happened when Colonel Drake drilled the well specifically to find oil. He essentially started the commercial oil industry. This procedure for finding and making oil available worldwide really took off late in the nineteenth century. So, since that time, we have been releasing extra energy over and above what the Sun supplies day in and day out. It is the energy of the entropy of the time. I have no reason to call this one catastrophe as entropy, but it was. The term "entropy" is a general term. I am using it in the specific sense just to give that particular unusable energy significance. I need to call it something. This word describes the gift we eventually received, fossil fuel. I will probably call it earthborn energy from here on.

Extra energy above the normal Sun's energy is not easy to come by. It requires a special kind of event and a long time to nurture before it can be recognized as an unusable energy that became useful. This is what transpired from 65 million years ago to the present time. The finding and use of oil was probably not considered an extra energy event. Many people may not consider that extra energy now. But why should this extra energy be a problem? For one thing, Earth is set up with a normal system of supplying the world its energy needs, the Sun. The release of this extra energy has been put to very good use. Its byproduct of carbon dioxide was not considered a major problem. At the time, we felt it was handled by the radiation of infrared energy out to space. The event of 65 million years ago is an abnormal one. Earth could not handle it back then. It may be having problems with handling what is left of it now.

My Conclusion

I have not found a relationship between the increase of carbon dioxide and any change in temperature or written evidence of an increase in the levels of the oceans due to any ice melting.

Up to about a billion years ago, ultraviolet energy from the Sun entered Earth and inserted much more energy than is inserted now on a daily basis. When Earth, through photosynthesis, provided plants, which then generated the free oxygen that was needed to provide an oxygen atmosphere; the planet itself altered the solar energy that it received. The Sun's ultraviolet rays split the oxygen, resulting in ozone in our atmosphere approximately a billion years ago. With ozone in the atmosphere, it reduced the amount of energy being received from the Sun by blocking the ultraviolet (UV), high-energy rays from inserting their energy into Earth. These rays contain the highest level of energy that can be emitted into Earth. On the electromagnetic spectrum graph, it shows that the wavelengths of light we see have energy of about 1 electron volt. The energy of the ultraviolet wavelength is approximately 100 times greater than the visible and 1,000 times greater than the infrared wavelengths.

About 500 million years later, the Cambrian Explosion occurred. Then we began to see terrestrial animals on Earth. Perhaps that event is the result of the energy of the ultraviolet being blocked. This condition remained until early in the twentieth century. Then, many new uses of synthetic gases were being developed. They were used for sprays and other commercial gadgets before eventually finding significant use before World War II in air-conditioning units for home and commercial uses. These fluorocarbons went up into the atmosphere and caused a hole in the ozone layer above Antarctica at the South Pole. This hole has allowed ultraviolet rays and their very high energy into our system, specifically in the region of the ice of Antarctica. Perhaps this is why we see some melting in the South Pole. Corrective action by the peoples of the world eliminated most of these gases. The hole is expected to be closed around the year 2030. Maybe we will see the South Pole stabilize then. The ice might return to a normal amount.

However, there is an interesting quirk here. Because the ultraviolet wavelengths produce much greater energy than any of the other wavelengths from our Sun, perhaps we should use that hole in the atmosphere above the South Pole to provide energy for us. If we could focus this energy onto some means of an improved solar cell, we might gain a terrific amount of energy from this unused source.

Interestingly, the ice at the South Pole may be influenced by the hole in the ozone layer. The ice of the North Pole might be the result of the entropy that provided us the energy of fossil fuels. I would say that the use of fossil fuels in the Northern Hemisphere is about ten times that of the Southern Hemisphere. Therefore, it could be impacting the Arctic ice pack. I have not seen evidence that the increase in carbon dioxide in the atmosphere relates to this point because I have not seen the temperature of Earth change very much. I have not even seen evidence that the level of carbon dioxide in the atmosphere is the same all over the world. It probably isn't. Radiation of infrared energy is probably not the same all over the world. Hawaii is a good constant source from one location, but it might not represent what is occurring all over the planet. It may not even represent what is happening with the climate in the Hawaiian Islands.

Essentially, I do not believe the increase in the atmospheric carbon dioxide to 380 parts per million is an indicator that we have an increase in store for us in the climate of the world. For all the generators of carbon dioxide that have occurred over the last 150 years, I find little evidence that it has caused any significant change in the atmosphere's concentration or any change in the weather, let alone a change in climate. Because Earth is in the act of recovering from a recent ice age, I believe we can expect the temperature to drift up. I believe the scientists are split on their concerns about a climate change. Because water vapor rather than carbon dioxide controls Earth's reradiation, I believe the stability of the atmosphere's water vapor is determining the current stability of Earth's climate. Water vapor is a by-product of Earth's warming by the Sun, which has been constant. Carbon dioxide is an earthborn supply of energy, which is limited. Even though water vapor changes dramatically from day to day, that is the part of our balancing system to prevent global temperature from running amuck. The massive inertia that the world's oceans provide also supplies a large sink for carbon dioxide, which remains an obstacle for any change in climate. Although I see no global warming from my study, I do find a more immediate problem that I will discuss later in this book.

I guess that most of all I don't accept the "Global Warming people's" placing the blame on the exhaust from automobiles for the increase of carbon dioxide in the atmosphere. Although automobiles do exhaust carbon dioxide that end up providing a significant amount into the atmosphere, it is rather small when you consider the amount of carbon dioxide placed in the atmosphere as a result of six and half billion people now being on earth. I have shown in this book that the infrastructure support required to provide man his daily Calories places ten times as much carbon dioxide in the atmosphere. The combination of man and this support system of animals to provide food for man add ten times the amount of carbon dioxide in the air as cars do. Here is an area that can be attacked with good results for reducing the carbon dioxide into the atmosphere, while at the same time reducing the amount of Calorie intake of food by the rich nations of the world. This reduction would result in better health as well as reducing any effects that carbon dioxide might have on the climate. Many people consume 3500 Calories per day. This should be cut to 2000 per day. The important strategy about this approach is that by reducing the world's intake of Calories it reduces the amount of animals needed in the food chain, so we would get a double impact from this. Not only would we reduce the carbon dioxide output of man, we would reduce the carbon dioxide output of the animals we don't need to keep around for food. You should review the chart that I have generated after researching for the number of people in the world since 500 BC.

Population growth compared to the increase in atmospheric Carbon Dioxide

Scientists are talking about the atmospheric carbon dioxide increasing rapidly. The increase from 1850 to 2004 was from 280 parts per million to 380 parts per million. This is about a 36% increase in this time span. It is a growth of 100 parts per million.

If you think that is a major increase, look at mankind's increase over time. Remember, each of these people walk around with a body temperature of 98.6 F (37 degrees Celsius) and breathing out carbon dioxide (about a pound a day). Also remember that the food to keep them alive requires significant numbers of animals and plants to keep them alive. So, the number of these animals is increasing at a slightly greater rate than man. Together, man and the animals required to provide man his food put out ten times the carbon dioxide as car exhausts do. See the data for the people growth. I will start with 500BC where there were 100 million people on earth. (52)

Year	Population in millions	Change Pop/years	Change/year
500 BC	100		
700 AD	200	100/1200	0.083
1500 AD	425	225/800	0.28
1800 AD	900	475/300	1.58
1850 AD	1200	300/50	6.00
1900	1625	425/50	8.50
1950	2500	875/50	17.5
1975	3900	1400/25	56.0
2004	6500	2600/29	89.7
2006	6700	200/2	100.0

So, from 1850 to 2004 the population went from 1200 million people to 6500 million in 2004. This is a growth of 442%. This is a percentage growth that is 13.4 times the growth of atmospheric carbon dioxide during that period. Notice the growth. You

remember the scientists and their graphs showing the sudden leap in atmospheric carbon dioxide indicating how much the trend was rapidly up. If I were to graph the curve for the change in population per year it would jump up more dramatically than that carbon dioxide curve. The jump on the graph from 1850 would be a jump of over 15 times compared to the carbon dioxide curve jumping up by a third. This is a 45 times differential for the population growth over the carbon dioxide growth.

Summary – Summary

With the increase in cars from zero to 650 million from 1850 to 2004 – which is an indeterminant percentage? With the population growing from 1200 million in 1850 to 6500 million in 2004, which is a percentage growth of 442% - each of which, when totaled together contribute the majority of the carbon dioxide (forgetting about volcanoes which probably contribute as much as both of these put together) I find it impossible for the atmospheric carbon dioxide to have only increased by 100 parts per million (36%). But, lets say it has increased by only 100 parts per million. Why? This means there is something that the scientists aren't taking into account that is offsetting the amount of growth that we should have seen in our atmosphere. Either the carbon sinks are doing their job quite well and eliminating most of the carbon being generated or the precipitation is taking the carbon out of the air and putting it into the oceans, or there is another phenomenon that is neutralizing the amount of carbon dioxide being generated by these two big generators. They are looking at two things, a slight increase in the temperature over a one hundred and fifty year period and an increase in the carbon dioxide in the atmosphere and trying to make a match out of them. Worse yet, they are saying what is causing this small increase in absolute value (100 parts per million) of carbon dioxide in the atmosphere is the automobile exhaust. And worse yet, these 100 parts per million are causing a climate change. As far as I can determine we are slowly coming out of a Mini-Ice Age. There is no way they can correlate the small growth in atmospheric carbon dioxide, during a huge increase in carbon dioxide exhausts from cars, people and the animals that people take for food to any global warming. Meanwhile something worse is happening and is discussed as we move further on in this book.

Are we missing the obvious?

We may be missing the obvious. Since the time when the second atmosphere formed on Earth and we began to have photosynthesis, plants grew. Where did this energy come from? Let's assume it came from the sun's light as we always do. However, the plants don't give it back. When man and other oxygen breathing animals began their appearance on Earth it started to work like a battery that gets charged. The plants that had used the Sun's light to produce food now saw the food being consumed by these humans and other animals. So, what we have here is the plants stored energy being transferred to man and the animal kingdom. This was not energy given back to space; it was stored in each of the humans and animals of the world. Now, as the population of these creatures increased, they kept taking the energy stored in the plants from the Sun's light and began their walking the face of the Earth with their 2000 Calories a day for man and much more in each animal. However, this cycle is not a balanced cycle since more and more animals and humans arrived on Earth, each became a carrier of the energy derived from the plants. Naturally, man began to grow food in around 300 BC when the population was around 100 million people and I don't know how many animals, but man was domesticating these animals and now their growth was tying up some additional energy. Keep this picture in mind; the plants take light and convert it into food energy and then man takes it and, let's say for the sake of simplicity, that each man then carries these 2000 Calories around with him for life, even though he has to replace it each day from the plants and animals. As the population grows, there are more and more of these storage sites for plant's food to be stored, not only in man, but in the incremental growth of the population of animals that man eats for food. So now the population from 300BC until this year has increased by six and a half billion humans and a bigger percentage of animal growth than the growth of man.

So, here we have 6.5 billion people and a large contingency of animals that weren't here 2300 years ago. In fact the number of people has grown tremendously as shown on the chart of man's growth since 1850. So, over this amount of time, man has actually perpetuated this system by growing huge amounts of food in the form of plants year in and year out to provide him his 2000 Calories a day to keep his battery charge. So, in the year 300BC when there were hardly any people or animals on Earth till today, we have stored this tremendous amount of energy in people and animals. These 6.5 billion people are walking around with a temperature of 98.6 degrees Fahrenheit (37 degrees Celsius) and animals with even a higher temperature and where there was no heat from this large contingency 2300 years ago, now there is 640 billion degrees Fahrenheit (71 billion degrees Celsius) of temperature walking around in the form of humans. There is at least this amount of thermal energy walking around in animals. This was not true 2300 years ago.

So, maybe the obvious I am relating to is the huge amount of thermal energy the Earth has taken on. When man took his first bite of plant food he began to be a human battery that would eventually be full charged and as he propagated his species, there were these additional batteries. It is a fact that man gives off these energies each day and returns them to his body through the consumption of plants and animals that eat plants, so I only count these 2000 Calories once. I didn't count the total accumulation he took in during his life time and burned off. Since I am only counting his 2000 Calories once, it is the Calories that are derived from the Sun through plants and it is the only energy on earth that is not returned to outer space as all other energy derived from the Sun does. This was shown in the graphs of the amount of energy obtained from the Sun and the amount that was absorbed, radiated or reflected back into space. All of these humans and animals represent stored energy that is not given back to space, only the amount burned off each day is given back to space. It's a fact that man dies and his 2000 Calories go with him, however, I am only counting the incremental growth of man. More are being born than are dieing. Since incremental growth is used in these calculations, there is no need to count the every day consumption of food and the giving off of heat in one form or another; or the death of an individual. Incrementally the Earth has seen a huge population growth of man and animals All of this being a fact, then we probably have an increase of about a degree in Earth's temperature due to this large amount of population growth of humans and animals on Earth. Maybe it's as simple as that. The population is actually up to 6.7 billion people on the last report I saw. The incremental growth since 1850 is phenomenal and could be a problem.

Smithsonian Research

Presently, several funded programs are reviewing the subject of global warming. There has been very little analysis of the areas in the Southern Hemisphere. Most of the present-day programs are in the Northern Hemisphere. In addition, the greatest industrial output in the world is in the Northern Hemisphere. We are the greatest generators of carbon dioxide exhausts. We have been studying the subject matter to see if there are means to reduce the problem. More recently, a grant has been finding the status of the tropical regions of the Southern Hemisphere.

The Smithsonian Tropical Research Institute (STRI) has received an $8 million grant from HSBC to fund the world's largest field experiment on the long-term effects of global change on forest dynamics. A new Global Earth Observatory system will compare climate change and forest carbon data from seventeen countries around the world.[76]

---

[76] En.wikipedia.org/wiki/World_energy_resources_and_consumption; Energy Information Administration, United States Department of Energy, "World Consumption of Primary Energy by Energy Type and Selected Country Groups, 1980–2004," July 31, 2006.

Beyond Global Warming

I have covered global warming from my perspective. Now I will cover the "beyond" portion of this book. The "beyond" comes about from what I found as I investigated the question on global warming. As I generated the information on the energy consumed by man in a day and then began reviewing the amount of infrastructure required to support man, I realized we do not have to fear global warming. There is a more important subject matter to be investigated. It is the need to find some other energy to replace oil in the immediate future. When there were approximately a billion people on Earth, they could survive with the energy that the Sun supplied to Earth around them. If they were cold, they could light a fire and keep warm. If they could not see at night, they would burn the "midnight oil," so to speak. They could work off the barter system and trade services with other people.

In those days, people lived a rough and rudimentary form of living. Fortunately, along came the industrial revolution. Without it, there was no way to provide the bare essentials for 6.5 billion people that live on Earth today. The initial beginning of the industrial revolution was built around the use of steam. The steam was generated by burning coal or wood. After the oil strike in Titusville, it opened another huge opportunity for man to invent ways to use the products of oil and natural gas in addition to coal. The beginning of the twentieth century was the beginning of the big use of oil and its products. The Wright Brothers, their flight, and the planes they developed after that would have been only a dream without gasoline. The automobiles that became the normal means of transportation during the 20th century would have remained as steam engine driven vehicles and the rather cumbersome means of transportation would have slowed down the civil growth of the world.

Now I look at the future and see a big hole in the ability of man to produce and deliver the foods needed for a much bigger population. In all the energy substitutes I have reviewed, there are good substitutes for the products that oil brought us in many areas, but none are in the area of a mobile use fuel; mobile use that is portable fuel that can be put to use in remote places, at high-energy levels. I see no replacement for gasoline or diesel fuel, nor a replacement for the vehicles using these fuels and the infrastructure to maintain everything.

I am not talking about automobiles, but they are in this loop. I am mainly talking about huge equipment that requires significant power from a mobile source of energy. I am talking about equipment that requires thousands of horsepower to operate. When one thinks about the ethanol and other by-products of corn or sugar as a substitute, I have a hard time seeing anything in this area of ethanol that can supply 150 million barrels a day. Can you imagine the amount of infrastructure around the world that would have to be put into place to service ethanol all over the world? Besides, you cannot get the energy from one gallon of ethanol that you can get from one gallon of gasoline. Therefore, the world would have to produce 200 to 250 million barrels a day to provide an equivalent energy. We would be robbing our food from our land. It's like stealing from Peter to pay Paul. Meanwhile overuse of the land causes the land to become depleted and less and less productive. Next, we would be looking for a way to live on less food or a food substitute.

I can see ways of providing energy for many of man's needs for stationary types of requirements. These requirements provide heat for the home and much of industry. These needs can be handled for a couple of hundred years by the use of coal, hydroelectric power, and atomic power through electricity and natural gas. I can see using solar power for home use and some industrial use. Unless a better means of collecting solar energy is found, it has a relatively low power use. I have patents on integrated circuits solar cells that increase the efficiency of present-day solar cells, but I do not see a way of providing the large power uses for heavy machinery away from a stationary supply of power. Most importantly, I see no immediate equipment or fuel to handle the large jobs that large machinery, such as Earth movers, heavy harvesting equipment, cranes, and many other huge pieces of machinery that run on gasoline or diesel fuel, provide.

Where will we find fuels to fly the airplanes? The airplane industry uses a fantastic amount of high energy fuel and I don't see where it's coming from in the not too distant future. Without the large equipment required to work the mid West's large fields of wheat, corn, barley and other essentials for the world's food supply, there would be little food to spread around the world. Without the huge farms for raising cattle, pigs, and chickens, there is no way to provide these food requirements for the world. Without the transportation to carry these foods and other essentials to support today's population, how do we get the food and other essentials to where they are needed? How do we harvest the crops of the central valley of California, which contain the fruits, vegetables, nuts, and rice to feed the world? How do we harvest the many other fruits and vegetables that come from Florida, Mexico, and South America? We may be able to farm and catch fish. How do we catch enough if we don't have power driven boats? How do we transport the ones we catch to the world's mouths before they become inedible? We need refrigeration or ice to keep them from spoiling. Making ice takes energy. Present-day ice makers in the world run off gas (or electricity) most of the time.

Each day, we take for granted that we, the people of the industrial world, can get up and eat breakfast, lunch, and dinner without pause. It is amazing when you consider the fact that, every hour of every day, people all over the world are producing, shipping, delivering, serving, and making sure your food gets to you. They need to deliver it to receive money so they can buy the food they need. Think about the enormous energy being directed to make this happen like clockwork. People drive little trucks to bring all the foods to the supermarkets, restaurants, or your home. It is done every day and all hours of the day. They check the shelves and decide what they need to bring tomorrow or next week. Think about industry that has a "just on time" type of production mode. This mode is bore by the big industrial giants that do not want to carry inventory. They want it delivered "just on time."

The activities to bring these essentials to each of us that are so fortunate; are incredible. Amazingly, these people doing this are having it done for them also. If not, they would be in trouble of existing. There is no "just on time" method to grow the plants and raise the animals.

Remember when I covered the energy requirements to sustain man and his 2,000 table Calorie per day needs? As a result, he needed ten times this number of table Calories tied up in the animals needed for food and the entire loop to provide his daily need. This did not count the energy needed to bring this food to his table. It did not count the airplanes flying the food or the large transport tankers carrying food. It did not count the railroads and their part in the supply. It did not even count the small vehicles that deliver the food to the homes. They all take time to produce and deliver, so there is ten times the daily need for man in the loop at all times to be able to provide the daily needs. This long loop requires an enormous amount of energy. Forget about the energy that man needs per day to live. It is small in comparison to the amount of energy needed to provide this supporting infrastructure.

You now see what the beyond is. The beyond is what happens when we run out of the energy of the entropy of 65 million years ago, that is, earthborn energy of fossil fuels that came from that catastrophic meteor crash and other earthborn energy that lay in the ground for many years that have allowed the world to live its life, reproduce, and still be able to supply the many. What we could do with a billion people on Earth years ago, we will not be able to do with today's 6.5 billion, let alone the 10 billion people, which is the expected growth over the next thirteen to twenty years.

How much energy came from the meteor crash of 65 million years ago?

It's important that I get this message across so I am going to put it in dark bold print. The meteor crash that hit earth 65 million years ago and how it compares to the amount of natural gas reserves, the petroleum reserves, the coal reserves, the energy from the Sun each day, the total fossil fuel reserves versus the amount of energy released from the formation of the Chicxulub Crater in the Yucatan Peninsula. Note; that crash created a little more than ten (10) times the energy left in all the fossil fuel reserves. Now you see how fortunate we were to receive this messenger from outer space. Also, consider that this crash doesn't count the fossil fuels that were also in the earth in addition to this crash.

We have been using up these fossil fuels for about one hundred and fifty years. Of course the amount used up in the last 50 years has been the big hit. We cannot continue to use this amount of energy per capita for the next thirty years. We will need to find a substitute energy source. More specifically we will need to find an energy source that is portable such as gasoline is portable and has the energy equivalent of gasoline per gallon used; otherwise there is no way of providing the food for the general population.

- 6.0×10^{21} J, the estimated energy contained in the world's <u>natural gas reserves as of 2003</u> = 1.43 x 10^{18} Calories

- 7.4×10^{21} J, the estimated energy contained in the world's <u>petroleum reserves as of 2003</u> = 1.77 x 10^{18} Calories

- 1.5×10^{22}J, the total energy from the Sun that strikes the face of the Earth each <u>day</u>[4] = 3.58 x 10^{18} Calories

- 2.6×10^{22} J, the estimated energy contained in the world's <u>coal</u> reserves <u>as of 2003</u> = 6.21 x 10^{18} Calories

- 3.9×10^{22} J, the estimated energy contained in the world's <u>fossil fuel</u> reserves <u>as of 2003</u> = 9.32 x 10^{18} Calories

- 5.0×10^{23} J, the approximate energy released in the formation of the <u>Chicxulub Crater</u> in the <u>Yucatán Peninsula[7]</u> when the meteor crashed into Earth some 65 million years ago. =1.19 x 10^{20} Calories

I hope this brings the message to you. We are using this energy up fast, and we are increasing its use.

Whatever we replace this oil and its derivitives with, we must also provide the equipment to use it and the infrastructure to provide it worldwide. So, we don't have many years to invent, so we better start right now.

I predict the world's oil derived fuels will essentially run out in thirty to forty years. It may take a little longer to completely run out, but what are available to be delivered in thirty to forty years will be from a greatly shrinking pool of these high energy fuels. The fuel needed to deliver them will be in short supply. It will be highly contested for. We will find the people that have what little fuel is left, keeping it for their own needs. This will completely crack the world trade we presently enjoy.

I am not trying to forecast doom. I am trying to figure out how to get around the problem. However, there are substitutes for the other functions that oil products presently supply.

The End of Oil and Oil By-products

Today, the world uses almost 100 million barrels of oil per day, or it will in the next few years. Some reports say there are a trillion (1×10^{12}) barrels of oil in the ground that is recoverable by today's methods.[77]

Actually, 100 million barrels a day is equal to 3.65×10^{10} barrels of oil per year. At a recoverable amount of oil of a trillion barrels, this equals 27.4 years of oil left. Let's assume that the price of oil rises and this provides funds for finding additional recoverable oil. Let's assume we recover two trillion barrels. (52)Then we would have 54 years of oil, but this is only if the demand stays at 100 million barrels a day. With the population growth to 10 billion people in the next thirteen to thirty years, the demand will increase by 50 percent to 150 million barrels a day or more; probably by the year 2020. This would reduce the above numbers to 18.1 years and 36.2 years. This is if there are the refineries to process this amount of oil to useable fuels. Keep in mind that oil refineries are expensive to build, about the same cost as nuclear atomic plants. Why would someone want to build additional refineries if they know that there will be limited return on their investment?

Now you know what the beyond is about. I have little doubt that

---

[77] www.offshore-environment.com/facts.html

the second scenario is the correct one of the population growth and the increased demand. Therefore, a thirty-year period is close to the one to worry about.

Scientists may be right that Earth's temperature might rise a degree or two in one hundred years. That is nothing compared to the loss of our extra earthborn energy we have been putting to use for the last 150 years. I doubt we can do anything about a change in climate and the temperature of Earth going up a degree in a century. But this would be a minor problem compared to what I am talking about here. I know we must do something about replacing our mobile energy source in the next twenty or twenty-five years. If we do not do something, the world is in for a crash of some sort. This crash alone would reduce the problem of Earth's temperature rising if it is due to our use of fossil fuels. With or without limited oil, the amount used per day will decrease and eliminate any problems that may be associated with a temperature rise due to the use of oil. I would expect that the use of oil will peak in about ten years. It is not because of demand. It is because the ability to discover enough to meet our demands will limit the world's use. Any global warming would be a "handled" problem by an act of nature. Of course, some of the substitutes for the oil are dirtier than oil products. We might have to worry about their related pollution problems. It is better to have pollution problems to worry about that can be solved versus not having enough food to keep you going. So, the more immediate problem for the world is to find other means of providing energy from other sources. In parallel we must continue to work on the pollution problems that can arrive due to the use of coal for some of our industries that presently use oil products. Anything we can do to reduce the atmosphere's pollution now will be needed to handle the problems brought forth by the increased use of coal.

Review of Energy supplies

The following are my own inputs on what I believe are the capabilities and the limitations of each source of energy. I believe my comments are correct but could be modified by scientists working on the various issues and new methods. However, I find that most methods will not provide a mobile source of power that is within 15% of gasoline.

Oil	• Oil and its derivatives are used locally and worldwide and in real time for various requirements.
	• It is mobile, that is, it can be picked up at many places and be used remotely. It can provide mobile power to huge equipment. Remotely means it can be easily transported in a car, tractor, or bus in the refined form of gasoline or diesel fuels in the vehicle where it is used.
	• A gallon of gasoline used in these remote places on a myriad of different pieces of equipment, large and small, provides more energy in this gallon than any other portable or mobile fuel. Any replacement seen to date cannot compare to it for the energy per weight.
	• There is a very limited source based on today's use and finite supply. As the supply runs out, it could cause major catastrophic situations to occur with a major problem in thirty years. This is my estimate

Electricity	• Electricity is used locally and worldwide and in real time for various requirements.
	• It is not mobile, that is, it cannot be used remotely away from its source. Of course it is presently supplied over large distances with the use of high voltage transmission lines and transformers, but it can't be used effectively as a portable source of high power for large machines that are moving. It can be used at very low levels in batteries in remote operations.
	• The supply is not as limited because it can be generated using continuous earthborn sources such as waterfalls, dams, atomic reactors and natural gas. It is also available in solar cells for low power applications.
	• Electric cars could be a help, but no large vehicles are capable of doing the major tasks done by gas-powered vehicles and machinery in remote locations.
	• New types of vehicles could possibly be developed for using highly mobile electrical power.
Chemical	• With chemical, there is a limited means of delivery,
	• and limited daily supplies.
	• It is mobile in limited amounts, but it is limited in capacity and capability.
	• There is a limited source for worldwide use.
	• There is limited mobile use due to limited means of transportability and size. The world simply cannot run on batteries.

Atomic Energy	• At this time, atomic energy provides limited capacity. It has great capability though and could have high capacity.
	• It is expensive and can be dangerous. However, except for one major exception, it has not proven to be dangerous.
	• It is not mobile, but it has great worldwide potential. It could replace the loss of oil for non-mobile uses.
	• We have time now to begin to increase the number of generation stations. Unique approaches could make this mobile and replace much of the functions due to the loss of oil. It may be possible to pipe and pump steam around.
	• A combination of this power and newly designed vehicles to run on electric power would be a solution. It would be somewhat like the submarines that use atomic power.
	• Could not be used to power airplanes because of weight issues using present techniques.
	• May be possible to convert directly from atomic into electric to charge batteries on a continuous basis. Would require invention.

Natural Gas	<ul><li>There is a limited supply of natural gas, but it is much higher than the supply of oil. But additional sources may be found of both.</li><li>Natural gas comes from the same sources as oil; fossil fuels. I believe it takes 6 million years for natural gas to be formed underground as we now receive much from this fossil fuel.</li><li>It has very limited mobility at present.</li><li>Its present use is not as high as that of oil. Countries in Southern Hemisphere do not need as much for heating due to milder weather conditions and density of people.</li><li>Worldwide use in quantity is presently available.</li><li>Natural gas in the form of methane can be developed to provide additional stationary power.</li><li>It is the cleanest of fossil fuels and ideal for use in providing heat in the homes.</li><li>It has limited transportability; however it can be liquefied and transported.</li><li>It could replace the use of oil in heating and some vehicles. Special natural gas cars would benefit. It is presently used on some transportation systems.</li><li>Its biggest impact would be use for electrical power plants.</li></ul>

Coal	• Coal is more abundant than oil. It has a lower consumption rate than oil and will last many more years. It can be used in some of the applications of oil. • Its main use is for heating, but it is used to generate electricity in most countries. • The biggest user is China. New methods are being used to bring it to new markets such as liquefying it. • It is the dirtiest of the fossil fuels, at least in its present form. • It is not mobile. It doesn't have the Caloric energy of gasoline per gallon when used in mobile type applications. • Applications being studied to see where it can replace gasoline, but not likely in vehicles. It could be used for increasing the electrical power produced.
Battery	• The battery has limited capacity and capability. • It is very mobile, but it is difficult to transport for high energy use. • There is a need to find better and more efficient methods of production. • It leaves behind "dead" batteries that must be made more reusable. There must be a better method to dispose of used batteries. Perhaps made of different material • The present approach is not capable of handling heavy jobs. • Electric cars would benefit transportation, but it is not enough power for large production machines to handle food production or fly airplanes. • We have to use electricity to charge them. Solar power can charge for small applications.

Solar Cells	• Solar cells have a remote capability if located in hot country, desert, maybe as I suggested putting large solar cells in Antarctica beneath the hole in the ozone to capture the high energy of ultraviolet rays. They provide 1000 times more energy than the standard sunshine per area of reception; i.e. smaller cells or much more power per given cell size as compared against Silicon Solar Cells. There is limited production to service worldwide potential demand.
	• It unusable in less sunny parts of the world. Ideal locations are already sitting on oil in the Middle East and have no urgency to hook up solar cells.
	• If installed properly, they can last for centuries.
	• It is expensive to install, but over the long haul it should prove economical and last long. PG&E in California is building a large 553-megawatt plant in the Mojave Desert that will heat steam energy at $0.22 per kilowatt-hour.
	• At present, it is not a substitute power for vehicles. It requires too big a cell to provide the horsepower needed. A 4000 watt system could provide a 5.35 horsepower system for a lawn mower but would be much too large in size.
	• Need invention whereby Solar Cells are located in the stratosphere where they could pick up a thousand times more power per unit area and then have a way to transmit the power to earth stations.

	• It would require some invention and worldwide support, but I believe the North and South Poles could be used to provide large amounts of solar energy. Each has a six month period of light 24 hours a day. We could use energy from the North Pole for half of the year and energy from the South Pole for the other half year. (I just invented that)
Sugar fuel[78]	• Sugar fuel (ethanol) has limited capacity, capability, and supply. • It is much less efficient than refined oil, and it is more expensive than gasoline. Derived from sugar it would provide a better competitor to gasoline than that provided from corn. There is a plant in Mexico that would be the cheapest approach but is in limited supply at best. • It requires stealing from one source of limited supply (food and water) to supply another. • It is mobile but requires a larger tank than gasoline. May be too large for most applications. • Depletion of the soil, used for growth, year by year limits its long-term use. • South America is ideal for sugar supply, as is Hawaii. Nebraska can be used for corn fuel. Maybe another source can be found by the chemists of the world. • It might be difficult to supply 200 to 250 million barrels a day and still have some for food. Probably impossible.

---

[78] I call it sugar fuel because it is able to supply a better quality of fuel for cars at present than that supplied by corn.

Steam	• Steam produces much of the electricity today. • The former workhorse could be dominant again if better methods of generating were found, such as using atomic power to heat the water. Today's technology can make steam power practical today for stationary power. • It is not mobile, but it could be but would require invention and safety issues must be solved. • It could be worldwide. • It is used now via atomic power. • The energy to heat water comes from coal. We have the coal to heat the water and make mobile, high-energy vehicles. • We would need better methods than were used on steam transportation 150 years ago. We have better technology and materials for handling. • May need a completely new means of transportation just like the locomotive trains did back in the 19th century.
Wind	• Installations are increasing. Windmills are presently used in special locations to generate electricity. • It could supply local energy needs. • It provides clean energy. • It is not expensive. Its cost is competitive with oil. • It is not mobile. • It does not have a high capacity. • Efficiency is questionable. • More countries need to review its use. Use to charge electrical vehicles. • There are limited places in the world where it would be feasible.

Thermal	• It is presently used in random areas. • It is not mobile. • There is a need to go deeper than oil. • There is a need to find better ways of reaching deep into the Earth. Maybe oil rigs could be used when oil runs out. The rigs could be refit for boring to thermal underground sites.
Waterpower From tides	• It does not include the use of waterfalls and dams. • It is not in use. • There is significant energy stored in the tides. There needs to be a method of capturing the tides of the oceans. It is known exactly when the tides are going to come in and go out. A water generator made like a turbine turns the generator with slip rings so that the turbine always turns in one direction to supply electricity so as to capture the tides coming and going out. • It is not mobile. There is an unknown amount of energy to be captured, but it is large. • Its main use would be to supply electricity. Only useable sites are on coastal areas. Hydroelectric plants running off of dams and waterfalls. This is an ideal source of stable electrical power. Excellent supplier of large power with very clean source and no problems to speak for air pollution. This is obviously not mobile but it is a huge source of electricity. Limited in locations since it needs a dam or waterfalls to provide it. The dams could be built as they have in the past but you can't just go and get waterfalls.

Summary of Energy Sources

Oil has one major advantage to supply the world's energy because it is easily transported to the world market. There is a high capacity as long as it is around. Oil has the greatest energy capability of all the fossil fuels. This energy capability being lost is the major problem to be faced in our near future. There is no mobile source of energy that can supply the heavy energy requirements for the heavy equipment we have developed over the last hundred years. A gallon of gas has a terrific amount of energy and to date there is nothing to compare to it. If there is there isn't enough of it or it can't go into the present applications.

A fantastic infrastructure has been built over the years to supply this energy. It is mobile, or portable. In other words, it can be used all around the world. It does not have to be used at its place of origin. In addition, its use by cars makes it accessible through the mobility of the car and the many gas stations. It is reliable. It is convenient. People normally carry a tank in their cars, and they also know when their cars are getting near empty. They can go to a gas station and fill the tank. Other than the shortage that OPEC caused in the early 1970s, there has been an ample supply for the growing world market. At this point in time, the derivatives of oil, that is, gasoline, diesel fuel, jet fuel, and kerosene, are the only mobile fuels that can be used wherever people go as long as there is a vehicle to consume it. Much work is done with instruments like chainsaws, lawn mowers, and, many others. This is not true of any of the other fossil fuels or source of energy. When gasoline is no longer around, there is no substitute for its convenient use and the power it can produce on huge vehicles for heavy loads. Maybe someone will invent a vehicle that runs on something as convenient to obtain as gasoline and with the same energy per gallon. This is not likely, but I would like to see the challenge being met.

There will be plenty of sources for supplying electricity. If an electric car for general use was available, it would help to solve the problem of more than 650 million cars in the world. The use of specially designed vehicles that run on electricity and can work heavy loads is quite feasible. Even if a long extension cord was needed, it would be one solution.

Think of a tractor. Imagine power being supplied by a specially designed extension cord that pivots when needed. Envision this tractor going up and down the rows of corn in the Midwest. There could be a way. However, for very high-energy needs, there is not enough power to handle some of the huge pieces of equipment now being used today. Now you know how powerful and helpful a gallon of gasoline is. The electricity would be supplied by the burning of coal or hydroelectric power. We would be able to use today's large transmission lines to connect to this system. Maybe someone will invent a wireless vehicle that receives its power via the airways.

Then there are the airplanes, the major consumers of fuel. How do we handle their need? I do not see anything on the horizon to handle the need of fuel for aircraft. It is possible that aircraft would receive a special priority over other vehicles such as cars, whereby they receive gasoline when it is down to selecting who gets the last of the gasoline. However, I see of no other energy source for aircraft that has enough power to handle an airplane without loading it down with weight. Atomic power may be possible, but it may be too heavy of a load because of the ancillary equipment needed to make it useful. It would have to be used to provide heat for steam that would drive the airplane and I believe this would be too heavy and not practical. Besides, using atomic power in an aircraft is very dangerous. After one accident, it would stop being used.

The use of natural gas is almost as easy as the use of gasoline. It is not mobile because it must be used at the source. But it is almost mobile because it is rather easy to pipe to one's home or other places of use. A huge infrastructure has been established. However, it cannot be carried around very easily. Natural gas does not have the energy of gasoline. This might limit its applications. Natural gas reserves are in a more plentiful supply than oil. It has a longer future because its use is not as high as that of oil products. However, when oil is no longer available, the use of natural gas will increase worldwide. That will probably reduce the world's years of reserves that are available. Reserves of natural gas are less obvious than oil. Thus, the suppliers, engineers, and scientists are vague about the total supply of reserves available. Purportedly, there is at least one hundred to two hundred years of supply remaining. Natural gas is mainly methane. A low-pressure gas, it is easy to transport by piping and other means of transportation. Natural gas is the cleanest source of the fossil fuels, and it is found in many areas of the world. Natural gas is mainly used for heating homes and providing the energy necessary to heat water for industrial applications. Natural gas is used in many countries of the world to generate electricity, which is even easier to transmit. When oil is gone, there will be heating problems in many parts of the world that have not installed natural gas for heating purposes. They will use different varieties of oil for heating. Because natural gas is methane, it is possible that dump sites in the world will be methane generators in the distant future. If one would find a way to more rapidly and cheaply convert garbage and dump waste material to useable methane, they would do the world a favor. Natural gas can now be shipped in liquid form. However, special equipment is necessary for its use. The tanks are heavy and inconvenient to carry around.

Coal reserves are very high. The biggest user of coal is China. Their use is increasing. Coal is probably the dirtiest of the fossil fuels. Society will have to find ways to clean its exhaust so as not to add more carbon dioxide to the atmosphere. With the demise of oil products, the use of coal will greatly increase. It is expected that, around 2030, the limited supply of oil products will place more pressure on countries to use domestic coal supplies. Many countries have not explored their country for coal because of the availability of cheap oil and gasoline. This will change with the price of oil increasing. The world's supply of coal is probably 200 years. (53)

The United States is one of the biggest sources of coal, but it has not pushed that availability with the low price of oil products and the dirty work it takes to make coal available. New methods for mining coal have come along. More will come in the near future. However, these new methods are for the recovery of known coal, not provide new sources of coal. Coal's major uses are in supplying heat and supplying electrical power. We could go back to steam using coal as the source of energy, resulting in better designs of steam engines.

Final Conclusion

I have gone back in time to show that man's very existence on Earth is due to the global warmth provided by the blanket of the greenhouse gases that have been around for centuries. Without these gases, man could not exist. Will these conditions continue to be friendly ones? If not, how long will it be before major issues occur? For all the increase in greenhouse gases being touted over the last fifty to a hundred years, I have not seen an adverse effect on the temperature, let alone a change in climate. It continues to supply global warmth. Is there evidence that we are about to experience a major increase in the global temperature? I think not if carbon dioxide is proven not to be the culprit. However, this is based on the relatively small change in temperature over the past century and a half. The critics that blame carbon dioxide as the potential for global warming somehow do not consider that water vapor has a bigger impact on greenhouse effects than carbon dioxide. The average water vapor has been very constant over time. This being a determining factor for Earth's climate rather than carbon dioxide will eliminate the concern about a climate change. At this juncture in time, I do not believe the carbon dioxide increase in the atmosphere has any impact on our climate. The data shows that Earth is recovering from a mini ice age. What we are seeing is normal warming from that mini ice age for several hundred years. Of course I brought up the subject of temperature gain due to population gain of man and the animals he uses for food. If this is the culprit, then continued growth of the population will result in continued increase in temperature.

There is very little evidence that anyone has monitored, on a real-time basis, the parts per million of the combination of water vapor and carbon dioxide in the atmosphere. It has not been shown if the climate has been impacted. Because we know the climate has not changed, we know the combination of the two has not reached any critical point to date. It probably will not because the Sun's energy determines the water vapor cycle and carbon dioxide is an earthborn source of energy. As long as the water vapor is the controlling factor, earthborn energy is not significant. The two should be monitored together because a very small decrease in water vapor could offset any bigger increase in carbon dioxide. However, water vapor changes from day to day, going from 1,000 parts per million to a range that is 10,000 parts per million. It may go as high as 50,000 parts per million at peak times. Meanwhile, the carbon dioxide level is around 380 parts per million. The carbon dioxide level has gone from approximately 280 parts per million to this high of 380 parts per million over approximately a one-hundred-year period. We have no real evidence that it has increased the temperature or the climate. Compared to today, what was the water vapor parts per million a hundred years ago? I believe the combined amount of these two absorbers has not changed much. If all that happens over the next hundred years is a similar increase in the parts per million and a similar increase in the temperature, then I believe it is a secondary issue. There has not been any evidence that we are seeing a climate change due to man or oil products. However, as I pointed out, just the number of people carrying around their 98.6 degrees Fahrenheit may be all that is required to show the slight temperature change on Earth. Mother Nature has her tricks. One appears to be the ability to ignore a change in carbon dioxide in the atmosphere by 100 parts per million.

I went back to the time that Earth began as a cloud of dust was to determine because I wanted to show how tough and resilient Mother Nature is when it comes to responding to issues that are trying to shut her down. During those 4.5 billion years, there have been many attacks on this Earth's system. First, the solar winds stripped the first atmosphere. However, during differentiation, Earth built a magnetic field to offset the solar wind. We were then able to build our second and third atmospheres. Differentiation is the result of the gravitational pull on Earth's early structure, which resulted in the different layers of Earth to be formed. This resulted in the layers stratifying into the heavy elements moving into the center core surrounded by the next heavy layer into a molten metal layer called the outer core, followed by the next layer of lighter elements called the mantle and on to the crust of this planet. This resulted in a molten outer core, which swirls around as a result of Earth's spinning at 1,000 miles per hour. This swirling created a magnetic field that deflected the sun's solar winds. Thus, the second atmosphere that was formed was protected from the solar wind. This second atmosphere was deprived of the oxygen exploding out of the volcanoes because the chemical priority of oxidation of all the elements on Earth's surface was given a first priority. So the original oxygen being expelled from the volcanoes reacted with iron in the rocks on the Earth's surface and deprived the atmosphere of this oxygen. During this time, Earth was building up the ocean levels, which was proven by fossils of water single-celled life of 3.5 billion years ago. Being deprived of oxygen meant the second atmosphere was composed of a huge amount of carbon dioxide and water vapor, which set off the first greenhouse effect. This heated Earth, but it also initiated the first photosynthesis of Earth. With carbon dioxide, water vapor, and sunlight, Earth had all it needed for photosynthesis. Earth responded by growing trees and plants all over the world and in the oceans. What came from this evolution on Earth? These plants used up the carbon dioxide and water vapor in the atmosphere while generating huge amounts of oxygen. Eventually, huge amounts of oxygen were in our atmosphere and very little carbon dioxide. If the carbon dioxide level exceeded 800 parts per million, some plants began to die. If the carbon dioxide drops below 50 parts per million, they all die. Then Mother Nature's next act began.

Now that there were significant amounts of Oxygen available, Oxygen-breathing sea life began to leave the oceans, demonstrated by the Cambrian Explosion, which found fossils of multi-cell advanced sea life leaving the oceans 543 million years ago. This evolved into large, oxygen-breathing and carbon dioxide-exhaling animals over the next few million years. These animals began to supply the carbon dioxide needed by the plants; this combination between the animal life and plant life began to be quite balanced and remained in balance as the animal kingdom grew over the last 550 million years.

However, several space parts were flying through the air from the big bang; and 65 million years ago, one hit Earth, which delivered a huge package of kinetic and potential energy to Earth. Except for our constant solar energy that we get day by day, the solar constant, this is the only time in the last billion years that any energy of significance came into Earth. The solar constant energy from our Sun continued providing Earth's daily budget of energy every day of our existence since Earth was about 100 million years old.

This extra external energy coming into Earth had to be accounted for. Energy cannot be destroyed or created in a closed system. Our closed system was the outer extremes of our atmosphere and the Sun's energy entering it. The sun has been losing energy, but it can continue for about 5 billion more years. Meanwhile, Earth radiates energy back that goes into space. The combination of the solar constant, Earth's energy budget, and Earth's radiation back into space resulted in the energy going out being equal to the energy coming in. We were in balance. Earth obtained no additional energy over the last 100 million years except for the large meteor 65 million years ago. This energy had to be accounted for.

We have been living for approximately 150 years with the oil, natural gas, and some coal that is being retrieved from the carcasses of the dinosaurs that were killed by that meteor 65 mya (million years ago). Additionally, there is some other material that was buried for millions of years from plants and trees and bog swamps being buried under earth for at least 6 million years (It is said that it takes at least 6 million years for organic material to be converted into fossil fuel. What happens when that gift runs out? We once handled the needs of people when there were a billion on Earth and no significant amounts of fossil fuels being used to help supplement their lives. What happens when we run out of oil and there are 10 billion on Earth around 2037?

My analysis shows that the energy to support man with the 2,000 table Calories he expends each day comes from plants or animals that eat the plants. Forget the energy that the man's body uses. It is returned to outer space in the form of heat or it does some form of work that takes up his energy and now is found in another form. He receives no direct energy from the Sun. It does not need to be analyzed any further than that. His energy is replaced by the amount of table Calories he takes in each day that are supplied by plants or animals that eat the plants. However, as I pointed out, the incremental growth of the population of man and animals he takes as food may be the prime reason for any increase in Earth's temperature. Thank goodness those plants only require carbon dioxide, water and the Sun's light. They are the converters of the Sun's energy into mans energy.

My analysis shows that these animal support groups that man uses for food use up ten times the table Calories that man consumes. The plants get their energy from the Sun and photosynthesis without eating up anything. The animals that provide man with food use up nine of the ten parts of the table Calories because they cannot be grown the day that they were consumed. Therefore, there is about ten times the amount of all these food substances being produced each day and are in the supply loop to provide man with his daily food requirements. I did not even include the large amount of energy that the transporting of this food requires. I did not take into account the energy required for aircraft to fly. Plants take anywhere from four months to more than a year to reach the level required for food. All of these, including plants and animals, are in the food chain in the world each and every day. The details in this book show how they parcel out. So, as the oil energy runs out I am looking for energy to handle this total food chain. This includes the production, feeding, seeding, fertilizing, supplying water, harvesting, transportation, mobile suppliers, distribution, cooking, ridding of garbage, and all the energy required to perform these functions.

Rather than putting all our priorities on solving what to do about any carbon dioxide increase in our atmosphere due to the use of fossil fuels,[79] I believe we should be putting higher priorities on finding a replacement for oil, which I believe will be gone in thirty years. This lack of oil energy may take care of the carbon dioxide problem all by itself – if there is a carbon dioxide problem and not myth. However, many of the substitutes would create more carbon dioxide than the by-products of oil do, especially coal, unless man finds a way to eliminate this problem. So, the loss of oil may impact global warming if the substitutes generate more carbon dioxide than oil products have and if it is carbon dioxide that actually affects the climate of Earth.

---

[79] We should keep working on reducing our dumping of carbon dioxide into the atmosphere. This is a very worthwhile venture, regardless if it has any impact on global warming.

We will need to find a way to produce and distribute the food required to feed 10 billion people on Earth by 2037. We will find a solution for most of the energy requirements for man's needs without oil. There are ample ways for providing energy for heating and air-conditioning and doing many of the functions we are used to. Most of the stationary machines that work off electrical power supplied by coal, hydroelectric power, or the conversion of natural gas into more uses for heating around the world. Atomic power would supply most of the needs and be a clean method.

I have reviewed the various approaches for supplying energy to replace the oil derivatives. I do not see any source of energy in its present form for providing a means of supplying the equivalent of 150 million barrels of oil a day that would be required in thirty years. This statement is based on present use and present-day upside predictions of oil reserves.

We must develop a clean substitute for mobile energy use to replace the fuel we will no longer have. Any oil that is available will be in short supply for the demand placed on it. We will need to replace the infrastructure that has been installed to provide a mobile source of energy for the present level of 650 million cars worldwide. This infrastructure should be able to deliver 100 million barrels of oil equivalents per day. That number will be 150 million barrels in the near future. This cannot be replaced by ethanol, the closest thing to a replacement for the fuel of today's vehicles. Because it is less efficient than oil, we would have to replace the present oil with 250 million barrels of ethanol per day. Are we kidding ourselves? This is the corn or sugar we need to provide our food energy. I can see us developing better and more efficient electric cars. These cars would receive their electricity from batteries that can be charged by electricity supplied by provided from Atomic Power or Hydroelectric sources.

I do not see us developing anything that could run huge equipment that is mobile, that is, the equipment that does all the heavy work in the world. Examples include cranes, bulldozers, and farming equipment. I do not see us supplying new types of machines to use this new type of energy in thirty years. It would require a war effort like the one generated in World War II.

Even if we develop a new energy source, we need to develop the equipment to run off it. My best option is to drop back to steam with cars. Locomotives would be designed better than they were a hundred years ago and could burn coal. Even electrical power could be used for portable steam engines to power vehicles. Solar panels could be built above the huge machines that do some of this work. This will help relieve the mobile issue. But there is no solar power that is designed to handle heavy loads on these mobile vehicles. They would have to remain in the Sun all day to charge the batteries. If not, they could only be run on sunny days. There would be a limit to the size and horsepower of the vehicle that could run on solar power.

We only get twenty-five to fifty watts of solar power per square meter from the 350 watts per square meter that the Sun supplies to Earth.[80] We need to develop solar power to provide at least 125 watts per square meter to allow those vehicles to do anything worthwhile. Then, if you had a roof of 10 square meters, you could develop 1,250 watts of power. That's not even two horsepower. Maybe you could have a roof of 100 square meters on the really big equipment. It could then develop 12,500 watts of power. That's equal to twenty horsepower. Simply stated, there are no large power sources available that can supply these types of requirements.

We should set a limit to food intake by each person per day. Let's say we set a limit of 1800 Calories per day. This is enough Calories for man to survive and do a normal amount of activity per day. Reducing this amount of intake would result in further reduction of the animals we raise for slaughter each year. Remember the reduction of animals is a major reduction in Calories they burn off and the amount of carbon dioxide they expel. Perhaps we could try to reduce our Calorie intake

---

[80] The Sun delivers 1,366 watts per square meter to the outer part of our atmosphere, but only 51 percent is delivered to Earth's surface. It only supplies it for about 25 percent of the day. As a result, the average is about 350 watts per square meter per day.

from via a reduction in the meat we eat and move toward the more conservative approach of taking more Calories from plants.

We should reduce the population growth in the world and limit the total amount to ten billion people. Why? The reason relates to the maximum amount of plant food that can be derived from the use of the Sun's energy. The Sun provides 1.512×10^{18} Calories to warm the land and only 0.023 % of the Suns energy is used for photosynthesis. The amount of energy being used each day to provide food for man is approximately 1×10^{15} Calories. With a population growth to ten billion this would increase this to approximately 3×10^{15} Calories per day. This doesn't include any of the energy to supply this food. The energy used by vehicles, cooking, air transportation, and the preparation of the food uses up considerable energy and takes it close to the amount that the Sun can budget for this venture. Remember a considerable amount of the Sun's energy is used each day to provide the world with water. This is another reason for limiting the amount of food intake and the population growth in the world. The animals we eat drink a large portion of water that is greater than mankind takes in. Reduction of meat intake will reduce the amount of water considerably. Reducing the population growth would held to keep the demand down. There is limited water in the world and the number of people should be limited to a maximum of ten billion people to keep water use manageable. Keep in mind that we must find a way to transport what water is available to the people in the world.

Eight Major Objectives to Accomplish

We have eight major things to accomplish:

1. Develop cleaner methods of energy consumption, reduce waste, and reduce harmful by-products.
2. Find a replacement fuel for oil that allows us to have a mobile energy source that is capable of running heavy machinery. Let's call it, for the lack of another name, OIL II.

3. Develop mobile equipment that can operate off of OILII.
4. Develop a distribution system that is capable of delivering the food requirements for man, using the vehicles that operate off OILII.
5. Keep hammering away with our technology to find other means of providing new sources of mobile energy that can handle heavy power requirements.
6. Find a unique fuel for aircraft. Maybe OILII?
7. Reduce the amount of Calorie intake by people to a maximum of 1800 Calories a day, while reducing the number of animals required for food intake per year. Both burn up Calories and emit carbon dioxide beyond those of automobiles. This would reduce the demand on energy.
8. Reduce the population growth to keep the population of the world under 10 billion, while reducing the Calorie intake. The world cannot support, with only Sun's dependable energy, a population over ten billion.

The sections of this book that showed Earth over the many years without man demonstrates Earth's fantastic ability to cope with major forces being applied externally and internally. Earth has evolved over the 4.5 billion years of its existence from a barren world to the present, enjoyable one. It has also evolved into a very stable world. The Sun continually supplies energy and Earth relieves the energy to the atmosphere. While Earth and its basic energy from the Sun has provided mankind with the means of providing food and water for the many, it has taken an earthborn energy to be able to complement Sun's energy and allow for means of taking the world's people to another level of living as well as to a much greater population.

The advent of man and animals and their growth in numbers has, in some ways, dissipated much of this extra energy as we utilized it to provide the world's enormous economical growth. We owe our transportation and almost every means of human endeavor to this earthborn energy. But, while becoming dependent on it as our major resource during this time of such dynamic growth, we have used a great deal of this earthborn energy. Now we must plan how to accommodate what has been achieved by developing some other means of earthborn energy. This book has shown how Earth has been able to make subtle changes that are correction factors for previous attempts at destroying the balances established. I have tried to show that the bigger immediate problem is the possibility that man might not be able to cope with the loss of oil as a mobile energy source, resulting in the loss of a major means of providing food and water to feed the growing population of Earth. Man can help Earth by working on both of these issues and at least doing our intellectual and physical best to relieve these issues.

I have presented a problem where we have to find another means of complementing the Sun's energy with an earthborn supply of energy to replace the amount of oil derivatives that will be lost. This must be done in the near future, but I may be wrong on the number of years before this problem hits. I say it is thirty years away. This is not a long time to develop an unknown source of energy and the means of applying it. This deserves major attention among the world's economic powers. We are only able to use about 50 percent of the Sun's energy at this time because much of it is reflected off clouds or Earth to outer space. A significant step to offsetting our problems would be to find a way to recover one-fifth of that which is lost each day. Much of it is lost before it gets past our stratosphere to Earth. If we could find a way to capture the ultraviolet rays with their significant amount of energy that are tied up in the ozone 30 miles up in the atmosphere, we would eliminate the problem almost completely. That is a source of energy that is as large as the energy we derive out of today's oil products, but this is not in man's plans today and may never be. This would require great creativity by someone. However, we have brilliant people on this planet. If a goal is set, we will reach it. It may not be the ultraviolet rays that are being caught up in the ozone, but our technology advancements have brought us this far and should carry us forward. Remember, one of the things I mentioned was to find a way of providing solar power off of the ozone hole in our atmosphere that results in the UV light hitting parts of the Antarctica. This would provide approximately ten times the power we receive off of present day solar cells that work off the weaker sunshine that hits earth. Once a goal is set, we would create a plan on how to reach it. We always have been able to do this. Man is ingenious when it comes to solving a problem that is defined well.

My final hope

My hope is that this book brings to the forefront the major problem that the industrialized world faces in the very near future. It is not meant to denigrate the energies being directed towards the reduction of carbon dioxide or other harmful gases being released to the atmosphere. It is meant to bring to the forefront the real problem the world faces in providing food to the world due to a lost of a mobile/portable source of energy that is capable of supplying enough power to handle present day activities. The demand for this type of energy will be very dramatic within thirty years. In addition, it becomes obvious as oil products become scarce within this time span it puts a great pressure on the world to remain civil. Wars are normally a result of a country being deprived of a great need and they extend themselves to obtain it in whatever way they can; likewise, wars are sometimes a result of protecting what one possesses.

John Durbin Husher

GLOSSARY OF ENERGY TERMS

Please make use of the following glossary of energy terms. I have converted many of the energy units to Calories to make it easier to relate to present-day issues. This can help those of you who have a technical background and others who just want to understand energy better. Your electric bills and gas bills arrive in various units for payment. I believe the world should work off of the large Calorie or table Calorie, as we call it. This method better relates to earthborn energy and the people of Earth's eating habits.

GLOSSARY OF ENERGIES

Kinetic Energy

Anything that is moving is energy and requires heat energy for the motion. Energy moving is called kinetic energy. When a person is running, it is kinetic energy. The runner burns table Calories to achieve this motion. However, in some cases, the motion may not be perceived because the movement is in an increase in the molecular motion within the material. This is the result of added energy in the form of heat that excites the atoms and molecules, thus raising their energy level. The heat added may be in the form of visible light, which contains various wavelengths. Each has different heat energy levels. Einstein's Photoelectric Effect, for which he was awarded the Nobel Prize, is a prime example of the energy of light. Photosynthesis is a form of energy that light of different wavelengths provide. Kinetic energy is equal to $E = \frac{1}{2} mv^2$, where m represents the mass of a system in kilograms and v represents the velocity in meters per second. With these known, the kinetic energy can be mathematically determined.

Potential energy

Potential energy is energy derived by its position on Earth. Whether the potential energy is increasing or decreasing, it comes about by the movement of the potential energy to a different position, that is, higher or lower in height. It took energy to move it to the higher height and this energy is converted to the material that was moved to the higher position. Thus, its potential energy changes, but the change in potential energy is the result of the movement by some force. Let's assume a person moves it. The person had to use energy to move it to the higher height. The moved object takes on this energy less the energy lost in friction by the person moving the object. If the movement is rapidly down in free fall, it gains kinetic energy over and above its potential energy. It releases much of both of these energies upon impact at the lower height. [81]

An example of potential energy changing to another form is the Niagara Falls. If a dam prevented the water from falling down the waterfalls, it would be contained and represent a potential energy. However, without any dam, the water falls from a given height with its potential energy and changes to kinetic energy during the fall while losing some of its potential energy due to its loss of height. It hits the water at the foot of the waterfalls with an energy release due to some loss of kinetic energy and loss of potential energy due to the lower height. In 1932 the Niagara Falls froze over and no water was released from the frozen falls. It remained there in potential energy. When it thawed and released the ice and water it was back in the business of supplying energy to the river below. This lower river, therefore, takes on this energy, causing it to flow rapidly. Some energy is lost to sound, the noise it makes as a result of the impact of the waterfalls upon the lower river. This causes the air movement that result in sound waves. The sound waves represent a form of energy in motion that is emanated as a result of the impact on the river. Some small amount of energy is lost due to friction of the water against the stone sides as the water pours down the steep incline, but it is gained back almost immediately by the water that follows to pick up the friction's heat while losing some of its energy due to friction as well. And this goes on. Since man arrived on the scene, much of the energy from the waterfalls is captured by turbines built by man to convert this energy to electrical energy. This is done fairly effectively and with fairly good efficiency.

Energy has to move to have some immediate release to another form, even if it is only its atoms or molecules moving more or less.

The absorption of heat energy may increase potential energy, resulting in increased molecular movement within the body. It always results in some form of energy transformation. It may only increase the temperature of the body, therefore increasing the activity of the atoms or molecules and increasing its total potential energy. Or, the added heat may cause some other form of energy transformation. One example is the heating of water to cause the surface to evaporate. Let's assume the body of water may just be a glass filled with water sitting on a 6-foot-high ladder that has a potential energy based on the mass of the water and glass, gravity, and the height of the ladder. Let's assume some heat is applied, resulting in some loss of the water through the gain of heat energy and the resulting evaporation. If enough heat is supplied, the water may continue to evaporate or may gain enough total heat to boil. The water will be lost through vaporization. If the heat continues to be transferred to this glass of water, it eventually will boil away. Only the glass will be left with its reduced potential energy. The water that boiled away as a result of gaining energy is carried to a point of condensation elsewhere. The movement of the water vapor was an example of energy in motion while the glass lost some of its potential energy.

The opposite is true, that is, the reduction of atomic or molecular activity is a loss of energy. This again requires reduction of movement through the loss of molecular energy due to the heat loss. It is in the negative direction when compared to gaining heat.

Mass Energy

Energy may just be in the form of a mass and its energy potential is derived from Einstein's equation ($E = mc^2$), where the c is a constant relating to the speed of light. However, this energy relates to the atomic level that the average person does not experience or is not aware of when they are experiencing this phenomenon. In most cases, the energy of the mass is unusable, such as a large rock, but it still has this energy capacity as a result of its mass and potential energy. It just is a matter of fact that mass energy of a rock is very difficult to generate or release.

It is therefore relegated to its position of potential energy. The Earth has a mass energy. It is 5.37 x 10^{41} joules or 1.283 x 10^{38} Calories. The Sun has a much greater mass energy.

Radioactive Energy

There is the energy of radioactive material, which is both potential and kinetic energy. The emissions are the kinetic energy. The remaining material is potential energy. This is also a direct example of Einstein's equation. Once Einstein related his famous equation of mass energy in 1905 (E=mc²), other scientists reviewed what had been thought of as the "free" energy derived from radium's radiation. However, when they measured the weight in a precise fashion, they found that the radium lost small amounts of mass. Further investigation showed that this loss of mass exactly corresponded to Einstein's equation. So, the energy was not free. Energy could not be created nor was it destroyed. The sunshine we receive from the Sun is due to radioactive fusion created in the middle of the Sun and its release of this energy.

Chemical Energy

Stored chemical energy can be capable of being released; for example, the energy stored in a battery. This energy remains relatively fixed until the energy of the battery is called upon. This energy is released when the battery is connected to a load. It is measured by the movement of electrons through conductors to the load in the form of current. The flow of the electrons through the conductor plus the transfer to the load results in the transfer of the energy to the load and the heat loss during this transaction. Again, movement must be made for energy to be consumed. Even if the battery is not used, it will lose its energy over time due to movement. A large amount of chemical energy is consumed as plants take on growth from chemicals in the ground or as chemicals are placed on the ground to provide energy for plants to grow faster or more abundantly. Dynamite is a form of stored chemical energy that can be released through an explosion.

Fossil fuels are a form of chemical energy. This energy is released via the burning of the fuels. This form of energy applies anthropogenic contamination of Earth's atmosphere via the release of chemical compounds that add undesirable gases to our atmosphere, such as carbon dioxide, nitrous oxide, and methane.

Food Energy

There is energy stored in plants derived from chemicals in Earth plus the sunlight, carbon dioxide, and water and related photosynthesis. This energy is derived from these sources. It can be measured in table Calories of energy. This energy remains in a stored condition until a person or animal consumes it for food and takes on this energy in the form of table Calories during consumption of the plant.

Mechanical Energy

There is mechanical energy related to the potential and kinetic energy that a mechanism possesses. Until released, this mechanical energy is only a potential energy source. In order to have the mechanism do its function, energy must be supplied through electricity or some other form of energy. Once there is movement, there is also energy being transferred and some loss in the inefficiencies of the mechanical movement due to friction. The mechanical energy is released by a force function supplied by the electrical power being supplied which is equal to V^2/I where the V is the voltage and the I is the electrical current. There is always the loss of energy due to heat in the current flow of the electrical power. This loss is I^2R where R is the resistance of the conductor of the current. Of course the loss is in both the mechanical and electrical systems is in the form of heat which is energy in another form.

This movement is released in the form of $E = \frac{1}{2} mv^2$, that is, kinetic energy related to movement through the v^2 (velocity squared) component of this equation. Also, the potential energy is released in the m (mass) potential energy portion of this equation to become kinetic energy. The machine does do some work which is a transfer of energy from the machine per unit time.

Thermal Energy

There is thermal energy determined by the heat content of a system or the heat emanating from a body or substance. This thermal energy is in the form of excited atoms or molecules in the substance. It is usually described as heat capacity. The temperature of the material is not a measurement of its heat capacity because temperature only denotes its condition. A large body can have the same temperature as a small body of the same material, but the energy of the two can be quite different. The larger the body of similar materials, the more its heat capacity at the given temperature and the more energy in the form of heat it can transfer.

Animal Energy

There is an animal energy, the stored or motion energy of an animal or human. Their movements require the use of thermal energy in one form or another in order to produce these movements. Their release of energy is due to the release of stored energy that can be measured in table Calories. Without apparent motion, some of this stored energy is consumed just in the act of living, which is always through the use of Caloric energy. This energy is lost through breathing, thinking, smelling, seeing, sweating, digesting, and many other internal functions of the body. The animal or human can replace or increase this energy via the digestion of food Calories. Energy can be reduced due to the activity of the animal or human and not replacing the energy through the intake of food Calories.

Gaseous Energy

There is gaseous energy that is energy related to the thermal capacity of a gas related to its temperature and the increased activity of its atoms and molecules. Different gases have different thermal energy under the same conditions. All gases are energy in motion. An increase in the energy is derived by an increase in its thermal properties, which increases the movement and capability of the gas. Even if the gas is in a closed container, it has motion of its molecules or atoms, which can gain energy through compression or direct heating of the container. Some gases are explosive. If ignited, they will give off a tremendous amount of stored energy.

Wind Energy

There is wind energy that is supplied by thermal differences from the equator, due to the Sun supplying more energy to the equator than to other parts of Earth, to the North and South Poles and the spinning energy and motion of Earth. This is really a special case of the gaseous energy mentioned previously. The Sun applies the energy and the differential between the Equator and the cold Poles causes the air to have a North South direction from the Equator. The spinning motion of Earth causes a change in direction. The winds' motion is a combination of the Sun's heat energy and Earth's kinetic energy to provide the wind energy, which is heat in motion. The winds' energy can be tremendous, as can be witnessed by the damage a hurricane can invoke. This type of energy does not result in anthropogenic contamination of Earth's atmosphere.

Solar Energy

There is solar energy; a form of energy stored in a semiconductor material like silicon and called Solar Cells, which gains energy via the photons of light which strike the material. The photons of light are converted to hole/electron pairs in the semiconductor, which are collected in the depletion regions of the semiconductor junctions. They are carried out of its conductors in the form of electrical current to provide electrical power to a load. Here, the energy of photons in the form of light is required to provide the energy and motion of the current through the wires to the load.

Other Forms of Energy

There are many other forms of energy, but each always requires heat energy to create, increase, and release motion. The Sun may be the source of all the energy on Earth, but motion is the release of all the energy on Earth.

Definition of Energy Terms

With the different forms of energy and the various levels over wide ranges of energy, different measurement units had to be defined that made sense within the various industries for their level of energy being utilized. In order to understand the concept of energy within each of the various sciences and industries, it is important to recognize certain terms and how they relate to the overall function of energy. These units or terms are such things as joules, Calories, watts, kilowatt-hours, British thermal units (BTUs), Therms, horsepower, and other units used to express the use of energy within the various industries.

If you are interested at all in energy, then you should make a copy of this index. It will come in handy.

Joule

The joule is the basic scientific term used to describe energy. In scientific terms, it is the energy required to move an electric charge of one coulomb through an electrical potential difference of one volt. If we moved the electric charge through a battery of 1.5 volts, it would require 1.5 joules of energy.

Watt

The watt is also the energy consumed to produce one joule continuously for one second, or one joule per second is called a watt. Because the joule and watt are too small a term for the power used in everyday life, the kilowatt-hour is used. The watt is not a true international technical term for energy because it is a joule per second. This is a power measurement. However, the joule and second are recognized international terms of measurement. The watt as a unit of power is now accepted internationally as the means of measurement of energy; so most Americans recognize the watt and not the joule.

Watt-hour

The watt-hour (Wh) is a unit of energy most commonly used on household electricity meters in the form of kilowatt-hour (kWh). A watt has units of energy per time. An hour is a convenient time rather than seconds. When multiplied together, they produce a unit of energy called a watt-hour. The watt-hour is derived from the multiplication of an international unit (SI) of power (watt) and a non-SI unit of time (hour). This was discussed internationally for some time before this term was accepted.

One watt-hour is the amount of energy expended by a one-watt load, such as a small light bulb drawing power for one hour. For reasons of convenience and intuition, laymen and utilities tend to use watt-hours to measure energy rather than joules. The watt-hour is usually used in electrical terms, but it is also used in the measurements of Sun's heat input to Earth.

Kilowatt-hour

One kilowatt-hour is equal to the use of 1,000 watts for an hour, or 1,000 joules per hour. The kilowatt-hour is commonly used to measure electrical and natural gas energy. Many electric utility companies use the kilowatt-hour for billing their customers.

If one used a 100-watt light bulb for lighting, it would provide 100 watt-hours in an hour. If it was left on for twenty-four hours, it would use 2,400 watt-hours in a day or 2.4 KWhr. If it was left on for a month, multiply these daily Kilowatt-hours by the number of days in the month. That would result in an electric bill in kilowatt-hours.

For a monthly electric bill, one might see a bill for the use of several hundred kilowatt-hours. This would be the accumulation of the kilowatt-hours per day that add up to that many kilowatt-hours accumulated over a month. This is much more convenient than using the term of joules per second of energy used for a month. For example, a bill for 200 kilowatt-hours in a month, if expressed in terms of joules, would be 720 million joules in a month on an electric bill. This is not a reasonable unit to use. People would not comprehend the impact of the use of a 100-watt light bulb or other electrical equipment in their homes. Of course, industrial use is so high in a month that they even have to use units that work well within their systems.

Electron volt/Mega electron volt

The electron volt (eV) appears as a unit of measure at the other end of the energy spectrum. Scientists Use the electron volt to measure and define energy at the low-energy portion of the energy spectrum. It is equivalent to the energy required to move one electron through a potential difference of one volt. Physicists have found that a more useful unit of energy is the mega electron volt (MeV), or 10^6 electron volts. As the charge on the electron is -1.6×10^{-19} coulombs, it means there are 6.25×10^{18} eV in 1 joule. Conversely, 1 eV is 1.6×10^{-19} joules.

Calorie

A calorie (small calorie) is a unit of measurement for energy. In most fields, the joule has replaced it as the international standard unit. However, the calorie has been in common use for the amount of food energy one consumes and amount of energy one expends over a period of time. Therefore, the calorie retained its identifying feature of a measure of energy. However, there is one major difference. There is a small calorie and a large kilocalorie, which is equivalent to 1,000 small calories. The calorie falls into two classes:

- The small calorie or gram calorie approximates the energy needed to increase the temperature of 1 gram of water by 1 degree Celsius (0.55 Fahrenheit) This is equal to 4.184 joules. The large calorie or kilocalorie equals 4,184 joules and approximates the energy needed to increase the temperature of 1 kilogram (2.2 pounds) of water by 1 degree C.

  (0.55 Fahrenheit). This is called the kilocalorie. In food content, it is marked by a large "C."

- This **large Calorie** is used in our definition of calories in food and food intake. It is equal to the kilocalorie and is equal to 4,184 joules. One consumes the large calorie via exercising or just breathing. A person replaces the loss of Calories by taking on food Calories. **This unit is used throughout the discussions in the book.**

British thermal unit (BTU)

The British thermal unit (BTU) describes the use of thermal energy. It is used globally in the power, steam generation, heating, and air-conditioning industries as a measure of power. Although it is in common use in these industries, in scientific use, the joule and kilojoule have replaced it.

In North America, the BTU describes the heat value, that is, energy content, of fuels. It also describes the power of heating and cooling systems, such as furnaces, stoves, barbecue grills, and air-conditioners, to state their energy capability. Through early use of this term and the large number associated with its consumption, it was called out as MBtu to represent 1,000 Btu. This can be confused as a million BTU, but it remained to represent 1000 Btu's because of this historical character. Many industries have gone to MMBtu to represent one million BTU.

A BTU is defined as the amount of heat required to raise the temperature of 1 pound of water by 1 degree Fahrenheit. 143 BTU are required to melt 1 pound of ice at 32 degrees Fahrenheit. As is the case with the table Calorie, several different definitions of the BTU exist, which are based on different water temperatures. Therefore, it may vary by up to 0.5 percent. The value in joules is centered on 1,055 joules (or 1055 watts), which equals 1 BTU. One BTU is equal to 0.252 kilocalories, or about a quarter of a table Calorie.

Summary

Since different activities use different terms to measure their use of energy, we must review all these terms and see how they relate. In the end, they are all the uses of energy and when expressed over a time period they express power. One may have to convert from one type of unit measurement to another, depending on what one is trying to achieve. The basic measurement is in joules, but they are inconvenient to use at times. I think you will enjoy seeing conversions of all these terms on a few sheets of paper, and it will provide you with a reference for everyday use.

CONVERSIONS OF ENERGY UNITS

Joules and Watts

1J = 1 joule = 1 watt for one second or watt-second = 1Ws

1 joule = amount of energy required to heat 1 gram of dry, cool air by 1 degree Celsius = 1W

1W = 1 watt = 1 joule per second

1MW = 1 megawatt = 10^6 watts = 10^6 joules per second

1 gigawatt = 1GW = 1 x10^9 watts

1 terawatt = 1 x 10^{12} watts

1 watt-hour = 3,600 joules = 3.6 x 10^3 joules = Wh = 3.413 BTU

1 kilowatt-hour = 1 kWh = 3.6 megajoules = 3.6 x 10^6 joules

1 gigawatt-hour = 1 GWh = 1 x 10^9 watt-hours = 3.6 x 10^{12} joules

1 terawatt-hour = 1 TWh = 1 x 10^{12} watt-hours = 3.6 x 10^{15} joules

1 petawatt -hour = 1 x 10^{15} watt-hours =

1 joule = 6.242 x 10^{18} electron volts (eV)

1 EV = 1 electron volt = 1.6 x 10^{-19} joules

1 MeV = 10^6 eV = 10^6 electron volts = 1.6 x 10^{-13} joules

1 GJ = one gigajoule = 26.8 m^3 of natural gas at a defined temperature and pressure

1 TJ = 1 terajoule = 10^{12} joules = 10^{12} watts = one trillion watts

1 horsepower = 746 watts

1 horsepower-hour = 2.686 MJh = 2.686 MWh

1 ton of TNT = 4.2 x 10^9 joules = 4.2 x 10^9 watts

1 kT TNT = 1 kiloton of TNT = 4.2 x 10^{12} joules = 4.2 x 10^{12} watts= 4.2 trillion watts
15–20 kT TNT = 63 x 10^{12} W to 84 x 10^{12} W = energy range of each of two atomic bombs dropped in World War II
1MT TNT = 1 megaton of TNT = 4.2 x 10^{15} joules = 4.2 x 10^{15} watts

Calories

1 calorie = 1 small calorie = 4.186 joules = 4.186 watts = energy to heat 1 gram of water by 1 degree Celsius
1 Calorie = 1 kilocalorie = 1 C = 1 table food calorie = 1,000 small calories = 4,1868 J = 4.1868 x 10^3 watts
1 C = kcal = energy to heat 1 kilogram of water by 1 degree Celsius
38 C = 38 food calories = energy released by metabolism of 1 gram of fat
2,000 Calories = average intake of food Calories per day of man = average output of energy per day by a man
2,000 Calories = 8.1868 x 10^6 watts of energy consumed and burned per day = 8.187 megawatts per day = 8.187 MW/day
1,500 Calories = 1,500 C = average intake of food Calories per day of woman = average energy output a day for a woman
1,500 Calories = 6.28 x 10^6 watts of energy consumed and burned per day = 6.28 megawatts per day = 6.28 MW/day

Converting Watts to Calories on Common Functions

Assume the average person consumes 1700 Calories per day and burns off 7.0 megawatts per day and there are 6.5 x 10^9 people in the world = 4.55 x 10^{16} watts/day = 10.9 x 10^{12} C per day=10.9 teraCalories per day = 10.9 TC per day
5.0 x 10^4 W = energy released by combustion of one gram of gasoline = 11.94 C
5.0 x 10^7 W = energy released by combustion of 1 kilogram (2.2 pounds) of gasoline = 11,940 C

200,000 W to 500,000 W = kinetic energy of a car at highway speeds = 47 to 119 C.

7.2×10^{10} W = energy consumed by the average automobile in the United States in 2000 = 17.197×10^6 Calories = 17.197 megaC

1.74×10^{17} W = total energy from the sun that hits the Earth in one second=41.6×10^{12} C

2.5×10^{17}W = energy release of Tsar Bomba, the largest nuclear weapon ever tested = 59.7×10^{12} C

1.04×10^{19} W = total energy from Sun that hits Earth in one minute=2.48×10^{15} Calories

1.339×10^{19} W = total production of electrical energy n the United States in 2001 = 3.198×10^{15} Calories

1.05×10^{20} W = energy consumed by the United States in one year (2001) = 25.1×10^{15} C

6.2×10^{20} W = energy from the sun that heats Earth in one hour =148.1×10^{15} C = 148.1 PetaCalories

1.5×10^{22} W = energy from the Sun that heats Earth in one day =3.58×10^{18}C per day = 3.58 ExaCalories

45.5×10^{15} W per day consumed and burned off by humans on Earth each day =10.9×10^{12} C per day =10.9 TC per day (based on 6.5 billion people)

6.0×10^{21} W = energy estimated natural gas reserves in world = 1433.0×10^{18} C = 1433 ExaCalories

7.4×10^{21} W= energy estimated petroleum reserves in world = 1767.5×10^{18} C =1,767.5 ExaCalories

BOE = Barrel of oil equivalent = 6.12×10^9 watts = 1.46×10^6 Calories = 1.46 MegaCalories

British thermal unit (BTU

1 BTU = 1,055 joules = 1,055 watts = 1.055 kW
1 BTU = 253 small calories = 0.253 kilocalories = 0.253 Calories
1 BTU = 778 ft. lbf (foot pounds of force)

1 MMBtu = 1 million BTU in natural gas = 1,000 cubic feet (Mcf) natural gas

1 watt-hour = 1Wh = 3.41 BTU/h

1,000 BTU/h ≈ 293 Wh

10,000 BTU/h = 2,930 Wh

1 horsepower = 2,540 BTU/hour = 746 Wh

12,000 BTU/h = 1 "ton of cooling" in air-conditioning = the amount of power needed to melt one short ton of ice in 24 hours

100,000 BTU = 1 therm (as used in heating and air-conditioning)

1 quad (energy) = 10^{15} BTU ≈ 1 exajoule $(1.055 \times 10^{18}$ J)[81]

---

[81] Quads are occasionally used in the United States for representing the annual energy consumption of large economies; for example, the American economy used 99.75 quads per year in 2005.

BOOK REFERENCES/CREDITS

(1)   En.Wikipedia.org/wiki/Carbon_dioxide_sink
      Carbon dioxide Falkowski, P.; Scholes, R.J.; Boyle, E.; Canadell, J.; Canfield, D.; Elser, J.; Gruber, N.; Hibbard, K.; Hogberg, P.; Linder, S.; Mackenzie, F. T.; Moore, B 3rd.; Pedersen, T.; Rosenthal, Y.; Seitzinger, S.; Smetacek V.; Steffen W. (2000). "The global carbon cycle: a test of our knowledge of Earth as a system". Science 290 (5490): 191-296

(2)   globalchange.umich.edu/globalchange1/current/lectures/first_billion_y... 4/28/2007

(3)   Geolor.com/geoteach/How_Did_Earths_Atmosphere_Evolve-geoteach

(4)   Geolor.com/geoteach/How_Did_EarthsAtmosphre_Evolve-geoteach

(5)   Search.eb.com/nobelprize/art-1367

(6)   Dev.nsta.org/ssc/moreinfo.asp?id=947

(7)   En.wikipedia.org/wiki/Photosynthesis
      Blankenship, R.E., 2002 Molecular Mechanisms of Photosynthesis. Blackwell Science.

(8)   Space.com/searchforlife/life_origins_001205.html

(9)   En.wikipedia.org/widi/Cambrian
      Gould, Stephen Jay; Wonderful Life: the Burgess Shale and the Nature of Life (New York: Norton, 1989)

(10) En.wikipedia.org/wiki/Oxygen

Campbell, Neil A.; Reece, Jane B. (2005). Biology, 7th Edition. San Francisco: Stanford University Press. ISBN 0-8047-1569-6

(11) en.wikipedia.org/wiki/Cambrian
Gould, Stephen Jay; Wonderful Life: the Burgess Shale and the Nature of Life (New York: Norton, 1989)

(12) en.wikipedia.org/wik/Oxygen
Campbell, Neil A.; Reece, Jane B. (2005). Biology, 7th Edition. San Francisco: Stanford University Press. ISBN 0-8047-1569-6
Dalrymple, G.B. (1991). The Age of the Earth. California: Stanford University Press. ISBN 0-8047-1569-6

(13) en.wikipedia.org/wiki/Oxygen
Campbell, Neil A.; Reece, Jane B. (2005). Biology, 7th Edition. San Francisco: Stanford University Press. ISBN 0-8047-1569-6

(14) en.wikipedia.org/wiki/Oxygen_cycle
Steve Nadis, The Cells That Rule the Seas, Scientific American, Nov. 2003 Cloud, P. and Gibor, A. 1970, The oxygen cycle, Scientific American, September, S. 110-123
Fasullo, J., Substitute Lectures for ATOC 3600: Principles of Climate, Lectures on the global oxygen cycle, http://paos.colorado.edu/~fasullo/pjw_class/oxygencycle.html

(15) en.wikipedia.org/wiki/origin of water on Earth
Jörn Müller, Harald Lesch (2003): Woher kommt das Wasser der Erde? - Urgaswolke oder Meteoriten. Chemie in unserer Zeit 37(4), pg. 242 – 246, ISSN 0009-2851

(16) en.wikipedia.org/wiki/Weather
O'Carroll, Cynthia M. (2001-10-18). Weather Forecasters May Look Sky-high For Answers. Goddard Space Flight Center (NASA

(17) math.ucr.edu/home/baez/temperature/

(18) aoml.noaa.gov/hrd/tcfaq/D1.html

(19) marshbunny.com/mbunny/sidetrin/hurricane/storms.html

(20) aoml.noaa.gov/hrd/tcfaq/D1.htm

(21) en.wikipedia.org/wiki/Tropical_cyclone
Atlantic Oceanographic and Meteorological Laboratory, Hurricane Research Division. Frequently Asked Questions: Why don't we try

to destroy tropical cyclones by nuking them? NOAA. Retrieved on 2006-07-25

National Oceanic & Atmospheric Administration (August 2001). NOAA Question of the Month: How much energy does a hurricane release? NOAA. Retrieved on 2006-03-31

© Copyright Commonwealth of Australia 2007, Bureau of Meteorology (ABN 92 637 533 532)

(22) interactive2.usgs.gov/faq/listfaqbycategory/get answer.aspp?id=187

(23) sec.noaa.gov/primer/primer.html

(24) en.wikipedia.org/wiki/Solar_radiation
      http://www.grida.no/climate/ipcc_tar/wg1/041.htm#121
      Construction of a Composite Total Solar Irradiance (TSI) Time Series from 1978 to present
      This article incorporates text from the Encyclopedia Britannica Eleventh Edition article "Sun", a publication now in the public domain

(25) en.wikipedia.org/wiki/Solar_radiation
      http://www.grida.no/climate/ipcc_tar/wg1/041.htm#121
      Construction of a Composite Total Solar Irradiance (TSI) Time Series from 1978 to present
      This article incorporates text from the Encyclopedia Britannica Eleventh Edition article "Sun", a publication now in the public domain

(26) okfirst.ocs.ou.edu/train/meteorology/EnergyBudget2.html

(27) physicalgeography.net/fundamentals/7j.html

(28) physicalgeography.net/fundamentals/7j.html

(29) en.wikipedia.org/wiki/Evaporation
      31 Sze, Simon Min. Semiconductor Devices: Physics and Technology. ISBN 0-471-33372-7. Has an especially detailed discussion of film deposition by evaporation
      Silberberg, Martin A. (2006). Chemistry, 4th edition, New York: McGraw-Hill, 431–434. ISBN 0-07-296439-1.

(30) climatesci.colorado.edu/2007/04/05/evaporation-is-equal-to-precipitation-on… 7/26/2007

(31) ajcn.org/cgi/content/full/78/3/660S

(32) Arizona Cooperative Extension, College of Agriculture, Tucson, Arizona

(33) upc-online,org/slaughter/2000slaughter_stats.html

(34) en.wikipedia.org/wiki/Aquaculture
Aquaculture Hepburn, J. 2002. Taking Aquaculture Seriously. Organic Farming, Winter 2002 © Soil Association
Timmons, M.B., Ebeling, J.M., Wheaton, F.W., Summerfelt, S.T., Vinci, B.J., 2002. Recirculating Aquaculture Systems: 2nd edition. Cayuga Aqua Ventures

(35) en.wikipedia.org/wiki/Mount_St._Helens
Mullineaux, D.R.; Crandell, D.R. (1981). The Eruptive History of Mount St. Helens, USGS Professional Paper 1250. Retrieved on October 28, 2006.

(36) math.ucr.edu/home/baez/temperature/

(37) from Joe Buchdahl, Mid-Holocene Thermal Maximum

(38) Smithsonian Release No reference found

(39) World-builders.org/lessons/less/biomes/SunEnergy.html

(40) En.wikipedia.org/wiki/World_energy_resources_and _ consumption
World Consumption of Primary Energy by Energy Type and Selected Country Groups, 1980-2004 (XLS). Energy Information Administration, U.S. Department of Energy (July 31, 2006). Retrieved on 2007-01-20.

(41) http://en.wikipedia.org/wiki/World_energy;_resources_and_ consumption

(42) www.offshore-environment.com/facts.html

(43) http://www.es.finders.edu.au/~mattom/IntroOc/notes/lecture04. html

(44) http://en.wikipedia.org/wiki/Earth

(45) http://links.baruch.sc.edu/scael/personals/pipb/lecture/lecture. html

(46) http://en.wikipedia.org/wiki/Earth

(47) http://en.wikipedia.org/wiki/Oxygen_cycle

(48) http://en.wikipedia.org/wiki/Ocean

(49) http//en.wikipedia.org/wiki/Ocean

(50) http://www.upc-online.org/slaughter/2000slaughter_stats.html

(51) http://www.avma.org.onlnews/javma/aug02/020801a.asp

(52) http://en.wikipedia.org/wiki/Sea_level

(53) http://www.msnbc.msn.com/id/5945678/

(54) http://wwwnationmaster.com/graph/ene_coa_pro-energy-coal-prod

http://www.csiforum.org/china.htm1

(55) http://en.wikipedia.org/wiki/Solar_variation

www.ingramcontent.com/pod-product-compliance
Lightning Source LLC
Chambersburg PA
CBHW032054020426
42335CB00011B/331